Lecture Notes in Electrical Engineering

Volume 346

About this Series

"Lecture Notes in Electrical Engineering (LNEE)" is a book series which reports the latest research and developments in Electrical Engineering, namely:

- Communication, Networks, and Information Theory
- Computer Engineering
- Signal, Image, Speech and Information Processing
- Circuits and Systems
- Bioengineering

LNEE publishes authored monographs and contributed volumes which present cutting edge research information as well as new perspectives on classical fields, while maintaining Springer's high standards of academic excellence. Also considered for publication are lecture materials, proceedings, and other related materials of exceptionally high quality and interest. The subject matter should be original and timely, reporting the latest research and developments in all areas of electrical engineering.

The audience for the books in LNEE consists of advanced level students, researchers, and industry professionals working at the forefront of their fields. Much like Springer's other Lecture Notes series, LNEE will be distributed through Springer's print and electronic publishing channels.

More information about this series at http://www.springer.com/series/7818

Khaled Khalaf · Vojkan Vidojkovic
Piet Wambacq · John R. Long

Data Transmission at Millimeter Waves

Exploiting the 60 GHz Band on Silicon

Khaled Khalaf
SSET-CSI
IMEC
Leuven
Belgium

Piet Wambacq
SSET-CSI
IMEC
Leuven
Belgium

Vojkan Vidojkovic
SSET-CSI
IMEC
Leuven
Belgium

John R. Long
Faculty of EEMCS
Delft University of Technology
CD Delft
The Netherlands

ISSN 1876-1100 ISSN 1876-1119 (electronic)
Lecture Notes in Electrical Engineering
ISBN 978-3-662-50947-0 ISBN 978-3-662-46938-5 (eBook)
DOI 10.1007/978-3-662-46938-5

Springer Heidelberg New York Dordrecht London

Softcover reprint of the hardcover 1st edition 2015

Printed on acid-free paper

Springer-Verlag GmbH Berlin Heidelberg is part of Springer Science+Business Media
(www.springer.com)

Acknowledgments

This work was fulfilled as a result of a cooperative research project between the TU Delft Electronics and IMEC wireless research groups. So, I would like to thank all those who helped me in bringing this work to life. I'd like to thank my advisors, Profs. John Long and Piet Wambacq, for their valuable guidance and discussions, and for making such cooperation possible. Special thanks to my direct supervisor, Vojkan Vidojkovic, for his constant support and continuous technical and un-technical discussions during my period at IMEC. I'd like to thank Bertrand Parvais, Kuba Raczkowski, Kristof Vaesen, Charlotte Soens, Viki Szortyka, Giovanni Mangraviti, Karen Scheir, and Julien Ryckaert for all their help, guidance, and fruitful discussions at IMEC. I'd like also to thank Roeland Vandebriel for the IO ring implementation. Thanks are extended to all who shared their thoughts and helped me in TU Delft including Yanyu Jin, Mina Danesh, Hakan Cetinkaya, and Antoon Frehe. I'd like to acknowledge all my professors at TU Delft including Leo de Vreede and Klaas Bult for their encouraging and valuable course contents, and all my classmates and friends in Delft who made it easier for me staying in that period just studying! I appreciate the efforts of all my friends in Leuven that were continuously supporting me in the good and bad times. I've to express deep thanks to my friends in Germany who actually encouraged me in continuing my post-graduate studies regardless of all the difficulties that I was facing. I'd like also to send deep regards to my bachelor staff members in Ain Shams University who guided me through such an interesting research and career path. I'd like to thank Alaa Medra for helping me finishing this book. No words can express acknowledging my family who did all what they can to help me achieving my goals, especially my wife and little kids for giving me a lot of "their" time to use in this book.

Contents

Figures

Tables

Abstract

Operation at millimeter-wave frequency, where up to 9 GHz of unlicensed bandwidth is available in the 60 GHz band, provides an opportunity to meet the higher data rate demands of wireless users. Advancements in silicon technology permit one to consider exploiting the 60 GHz band for commercial applications (e.g., short range, wireless HDTV transmission) for the benefit of end users. This could enable, for example, wireless streaming of uncompressed high-quality video packets of a movie in a few seconds due to data rates as high as multi gigabits per second. In this book, the design of a receiver front-end circuit for operation in the 60 GHz range in 90 nm CMOS is described. The following chapters include design details of the blocks used in the receiver, including: quadrature voltage-controlled oscillator (QVCO), local oscillator (LO) buffers, divider chain, low-noise amplifier (LNA), and mixer. The QVCO predicts 56.8–64.8 GHz tuning range from schematic simulations. The divider chain has 15 GHz locking range at rail-to-rail (0.5 V-peak) input signal. The LNA and mixer combination achieves a maximum conversion gain of 26.8 dB and a noise figure of 5.9 dB. The output −1 dB compression point is +6.3 dBm and IIP3 is −8.6 dBm. The complete front-end consumes 91.7 mW from 1 V supply. Physical layout of the test circuit and post-layout simulations for the implementation of a test chip including the QVCO and the first stage divider are also presented. Post-layout simulations show a maximum phase noise of −97.4 dBc/Hz over 55.4–61.7 GHz tuning range.

Chapter 1
Introduction

The unlicensed band centered at 60 GHz lies in the extremely high frequency (EHF) band, which is the highest radio frequency band[1] according to the International Telecommunication Union (ITU) running the range between 30 and 300 GHz [1]. This frequency range is equivalent to wavelengths between 10 and 1 mm in free space. That's why it's also called the millimeter-wave (mm-wave) band. The Institute of Electrical and Electronics Engineers (IEEE) has another frequency band nomenclature that assigns 60 GHz to the V band. The V band includes frequencies ranging from 40 to 75 GHz [2].

1.1 Motivation

The increasing demands of society for technology driven appliances is pushing the trend to shift operation to higher frequencies, and advancements in silicon technology is making this shift feasible. Data transmission is the current example of our choice. Almost no person can imagine carrying a laptop or any other portable device which is not connected to the internet, or even to a local network, from which you're transmitting and receiving information. These can be ranging from simple text information to streaming video data that requires large data rates of few gigabits per second. A movie, for example, can be steamed with more quality if uncompressed data is used. This needs larger amount of data, which can be transferred at the same speed, or even faster, using higher data rates. Higher data rates require more bandwidth, which is available at higher frequencies. Thus, operation at mm-wave frequencies is a good choice, as compared to lower frequency bands (e.g., WiFi MIMO at 2.4 or 5 GHz), to meet the current higher data rate demands of applications.

An unlicensed band of 7 GHz around 60 GHz from 57 to 64 GHz was assigned by the Federal Communication Commission (FCC) in the United States [3]. Frequency bands of 57–66 GHz and 59–66 GHz were also assigned in Europe and

[1]EHF is an International Telecommunication Union (ITU) radio band symbol. It is also equivalent to the ITU radio band number 11, which is the highest till the time of writing this book.

K. Khalaf et al., *Data Transmission at Millimeter Waves*,
Lecture Notes in Electrical Engineering 346, DOI 10.1007/978-3-662-46938-5_1

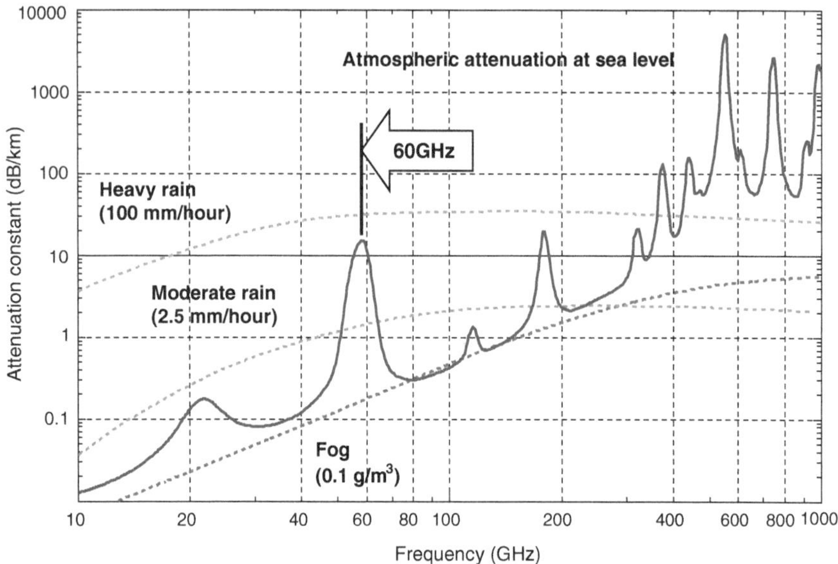

Fig. 1.1 Atmospheric propagation attenuation versus frequency [20]

Japan, respectively [4]. This is now encouraging commercial applications in the 60 GHz band. Natural spatial isolation caused by propagation loss due to oxygen absorption makes communication in this frequency band only viable over short ranges (till 10 m). Figure 1.1 illustrates the oxygen absorption peak in the 60 GHz region.

1.2 60 GHz Area Background

More information about the 60 GHz area from the system point of view is important to have a good background before starting circuit design. Circuit specifications were given as an input, and no system specs were derived from the standard. Thus, only some background information is going to be discussed in this section.

1.2.1 Standards and Frequency Plan

60 GHz frequency planning is covered in detail in both IEEE 802.15.3c [5] and ECMA-387 [6] standards.[2] Three modes of operation are defined in the IEEE

[2]ECMA is more related to European standardization.

standard: single carrier (CS), high-speed interface (HSI) and audio-video (AV). The ECMA standard also defines three devices: Type A, Type B and Type C. These modes or devices differ in their capabilities and performance with a variety of modulation schemes, different data rates (reaching a maximum of around 25 Gbps) and different operating distances reaching 10 m using line of sight (LOS) communication. The whole band (8.64 GHz) is divided into four channels with 2.16 GHz each, as depicted in Table 1.1 and graphically illustrated in Fig. 1.2. Adjacent channels can be bonded together to allow more bandwidth. Several possibilities for bonding multiple, adjacent channels exist for increased data rate.

1.2.2 Beamforming and System Architecture

Oxygen absorption at 60 GHz causes signal attenuation due to propagation loss. One advantage of the implicit attenuation for operation at 60 GHz is that signals cannot propagate more than 10 m and cannot penetrate walls. This increases security between two close offices for example. Directional propagation is used to enhance signal transmission and reception. In the transmitter, radiated power is concentrated towards the receiver instead of being wasted in unwanted directions. Similarly, gain is boosted in one direction and unwanted interferers can be spatially attenuated in the receiver. This suggests using multiple antennas at the transmitter, to direct and enhance signal transmission, and at the receiver, to improve the sensitivity and reduce interference. The size of an antenna is inversely proportional to the operating frequency. For example, 60 GHz operation allows the use of 16-element antenna array that occupies the same area as a dipole antenna at 5 GHz [7].

Table 1.1 Channels defined by the IEEE and ECMA standards

Channel ID	Lower frequency GHz	Center frequency GHz	Upper frequency GHz
1	57.240	58.320	59.400
2	59.400	60.480	61.560
3	61.560	62.640	63.720
4	63.720	64.800	65.880

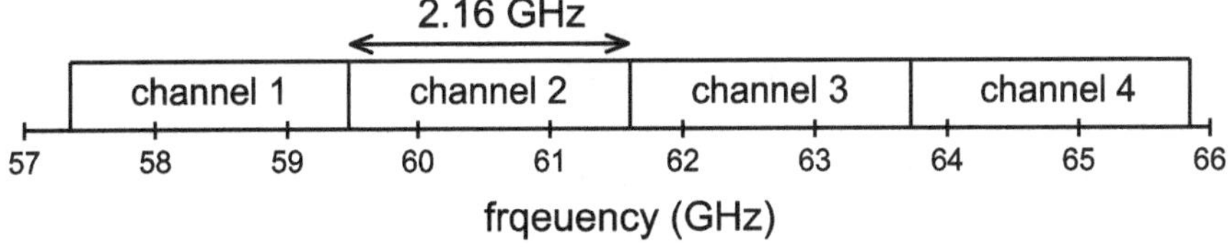

Fig. 1.2 57–66 GHz band divided into 4 channels [6]

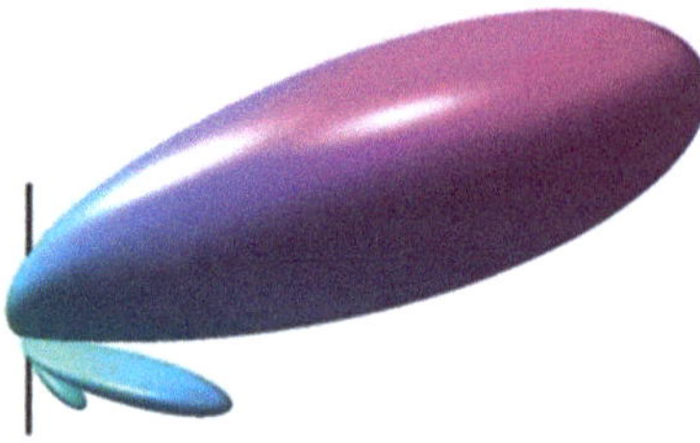

Fig. 1.3 Radiation pattern of a beamformer [21]

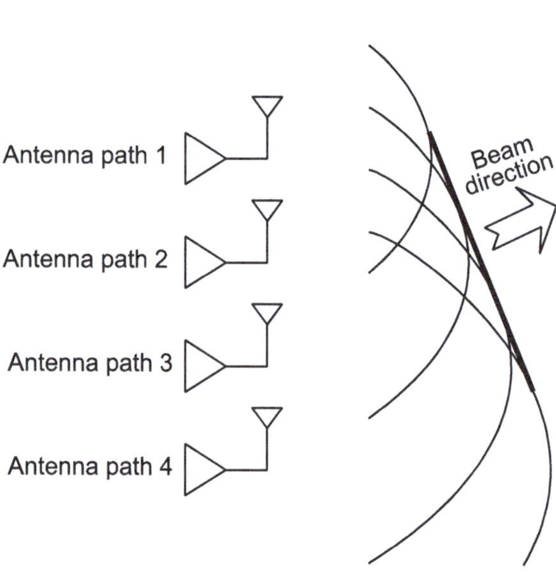

Fig. 1.4 Beamforming system with antenna arrays and transceivers

Beamforming is a signal processing technique used in sensor arrays for directional signal transmission or reception [8]. The term Beamforming is derived from spatial filters that were designed to form pencil beams (Fig. 1.3) [9]. As shown in Fig. 1.4, an array of antennas with variable gain and phase shifting (or time delay elements) can form different antenna patterns, one of which is a beam with a specific radiation angle θ [7]. Time delays among different antenna paths need to be compensated by true time delay elements for coherent signal combination [10]. Assuming no channel bonding, signal bandwidth is around 2 GHz, which is very small compared to the 60 GHz carrier frequency. In narrowband systems, phase shift blocks can be used instead of true time delay elements as an approximation for the multi-path signal to add constructively [10].

Phase shifting in a receiver can be implemented in four different ways: at RF after the LNA, at baseband after the mixer, in the LO path or using signal processing in the digital domain [11]. Signal combination in the digital domain uses the most hardware and is the most power hungry because the whole front-end is copied

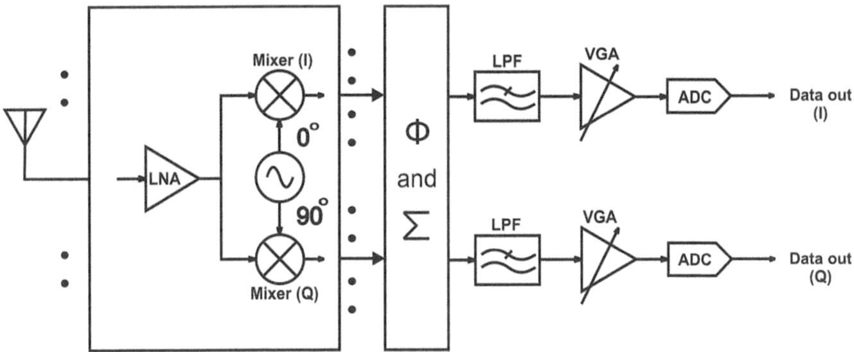

Fig. 1.5 60 GHz receiver architecture

as many times as the number of antenna paths. Phase shifting at RF (e.g., [12, 13]) places lossy elements directly after the LNA which degrades the system gain, noise figure and bandwidth. System gain and noise figure are less sensitive to amplitude variations in the large LO signal. Thus, phase shifting at LO provides the lowest effect on signal quality. Phase shifting after the mixer causes insignificant deterioration of gain and noise figure (as compared to phase shifting at RF). Signal combination is performed at baseband in both LO and baseband phase shifting. In both cases, in order to avoid using multiple PLLs, LO signal should be distributed to different antenna paths. This includes other problems, such as cross-talk and low LO power levels.

One antenna path of the receiver is shown in the block diagram of Fig. 1.5. Phase shifting and signal-combination are performed at baseband. The receiver is a direct conversion receiver, which includes a QVCO, LNA and mixer in the font-end. Antenna and baseband circuit design details are outside the scope of this book.

1.2.3 Enabling Technology

In a mixed-signal chip that includes analog and digital circuits, CMOS technology is preferred over bipolar for high volume applications when the digital part dominates. Moore's law states that on-chip density of transistors doubles every two years. This doubling is due to the fabrication of transistors with smaller minimum length. Smaller size transistors enable operation at higher frequencies. That's the reason for which operation at mm-wave became possible nowadays after it was just a dream years ago.

Scaling also has drawbacks. Smaller transistors usually require lower supply voltage due to the lower gate oxide. For example, the breakdown voltage is 1.8 V for 0.18 μm devices and 1.2 V for 0.13 μm ones [14]. This reduces the available voltage headroom, and thus, decreases voltage swings. The reduced supply voltage also limits the number of stacked transistors between supply and ground terminals. Smaller size transistors have more mismatch. This is because transistor mismatch is inversely proportional to the square root of its area [15].

Although devices are smaller in size, allowing for higher frequency operation, interconnects are not scalable as a consequence. Taking 60 GHz as an example, wavelength in free space is 5 mm. Assuming that the effective dielectric constant of a microstrip line is 4, the on-chip wavelength at 60 GHz becomes 2.5 mm. This means that a track length of more than 250 μm carrying a signal with frequency components of 60 GHz will cause a considerable difference in signal characteristics. This suggests the use of electromagnetic wave simulators, such as Agilent ADS [16] or Ansoft HFSS [17], to model relatively long interconnects, or with the help of a parasitic extraction tools, such as Mentor Graphics Caliber [18] or Cadence Assura [19], for medium and short interconnects.

1.2.4 Applications

The large bandwidth allocated for the 60 GHz frequency band could be used to transfer tens of gigabytes of data in few seconds. Short range indoor applications like broadband internet access and high speed point-to-point wireless communication could utilize this capability. Wireless replacement of the cable HDMI technology is another potential application that will change the dark picture of wireless video steaming in our minds. Figure 1.6 gathers the main applications, such as HDTV-DVD wireless connection, high speed mobile internet connection, wireless docking station and wireless digital video cameras. We can also go further for the high data rate connection applications, especially the ability to replace any short range cable connection with very a high speed wireless link. That's all of course beside the implicit security and isolation causing no wall penetration for the 60 GHz waves.

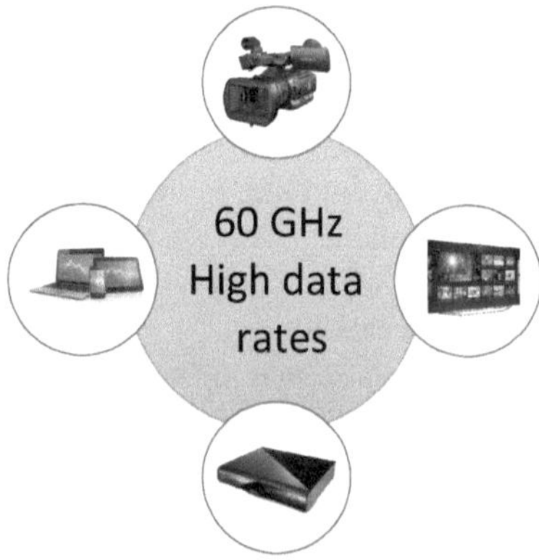

Fig. 1.6 60 GHz potential indoor applications

1.3 Book Objectives

The main objective of this work is to build a receiver front-end that can be part of a complete 60 GHz transceiver system. The front-end circuit includes a 60 GHz QVCO that drives a divider chain and a LNA-mixer combination through LO buffers. The QVCO should provide a phase noise < −90 dBc/Hz while oscillating at 60 GHz with 8 GHz tuning range. The divider chain is four consecutive divide-by-two blocks with locking range >8 GHz around 60 GHz at the delivered input power level from the LO buffer. The LNA and mixer combination should be reused from a design in 45 nm process. In the current 90 nm process, it should still provide a conversion gain >26 dB, noise figure <6 dB and an output −1 dB compression point > + 3.5 dBm. A subsystem including the QVCO with a LO buffer and the first stage of the divider chain is going to be presented separately with its layout.

1.4 Organization of the Following Chapters

The Book consists of six chapters. Chapter 2 provides a theoretical background on the blocks used in the design in separate subsections. Some basic concepts and then a summary of topologies from a survey of the recent literature are shown. A detailed design procedure with schematics and simulation results for each circuit block are explained and elaborated in Chap. 3. It also provides some information on the technology used, such as transistor f_T and the design of passive inductors, transformers and transmission lines. Chapter 4 connects all the blocks together showing the predicted performance of both the entire front-end and the QVCO + divider subsystem at the top-level with simulation results. Chapter 5 shows the whole physical layout and some detailed block layouts. It also provides post-layout simulations and comparison with the pre-layout simulation results. Some simulations for process, supply and temperature variations are also included in the chapter. Finally, conclusions and recommendations for future research are listed in Chap. 6.

References

1. International Telecommunication Union [Online]. http://www.itu.int/
2. IEEE Std 521-2002 (2003) IEEE Standard letter designations for radar-frequency bands
3. FCC (2001) Operation within the band 57–64 GHz, Code of Federal Regulations, title 47, vol 1, part 15.255c, 1 Oct 2001 [Online]. http://frwebgate.access.gpo.gov/cgi-bin/get-cfr.cgi?TITLE=47&PART=15&SECTION=255&YEAR=2001&TYPE=TEXT
4. Guo N, Qiu RC, Mo SS, Takahashi K (2007) 60-GHz millimeter-wave radio: principle, technology, and new results. EURASIP J Wireless Commun Networking, ID 68253 2007:48
5. IEEE Std 802.15.3c-2009 (2009) Wireless MAC and PHY specifications for high rate WPANs

6. Standard ECMA-387 (2008) High rate 60 GHz PHY, MAC and HDMI PAL. ECMA International
7. Niknejad AM, Hashemi H (2008) mm-Wave silicon technology: 60 GHz and beyond. Springer, Berlin
8. Wikipedia Beamforming [Online]. http://en.wikipedia.org/wiki/Beamforming
9. Van Veen BD, Buckley KM (1988) Beamforming: a versatile approach to spatial filtering. IEEE ASSP Mag 5(2):4–24
10. Zhao D (2009) 60 GHz beamforming transmitter design for pulse doppler radar. M.Sc. thesis, Delft University of Technology
11. Scheir K (2009) CMOS building blocks for 60 GHz phased-array receivers. Ph.D. thesis, Vrije Universiteit Brussel
12. Natarajan A, Floyd B, Hajimiri A (2007) A bidirectional RF-combining 60 GHz phased-array front-end. In: IEEE ISSCC digest of technical papers, pp 202–597
13. Roderick J, Krishnaswamy H, Newton K, Hashemi H (2006) Silicon-based ultra-wideband beam-forming. IEEE J Solid-State Circ 41(8):1726–1739
14. John R (2009) Long, RF integrated circuit design. TU-Delft M.Sc. course lecture notes
15. Pelgrom MJM, Duinmaijer ACJ, Welbers APG (1989) Matching properties of MOS transistors. IEEE J Solid-State Circ 24(5):1433–1440
16. Advanced Design System (ADS), Agilent Technologies [Online]. http://www.agilent.com/find/eesof-ads
17. HFSS, Ansoft [Online]. http://www.ansoft.com/products/hf/hfss/
18. Software for IC Design and Circuit Design Verification, Mentor Graphics [Online]. http://www.mentor.com/products/ic_nanometer_design/
19. Assura Physical Verification. [Online]. http://www.cadence.com/products/mfg/apv/
20. Kivisaari E (2003) 60 GHz MMW applications, Helsinki University of Technology, telecommunications and multimedia laboratory, 9 April 2003 [Online]. http://www.tml.tkk.fi/Opinnot/T-109.551/2003/kalvot/60GHz.ppt
21. Bass J, Rodriguez E, Finnigan J, McPheeters C (2005) Beamforming basics, connexions web site, 27 July 2005 [Online]. http://cnx.org/content/m12563/latest/

Chapter 2
Background

This chapter is going to present the theoretical background needed for the rest of the book. The blocks used in the design are going to be considered. This includes the quadrature voltage-controlled oscillator (QVCO), local oscillator (LO) buffer, injection-locked and static frequency dividers, low-noise amplifier (LNA) and the mixer.

2.1 QVCO

This section includes an introduction to the QVCO. The oscillation condition and main parameters defined in the cross-coupled LC oscillators are going to be reviewed. An overview on the origins of phase noise in differential LC oscillators is presented. Finally, the QVCO topologies discussed in recent literature are shown.

2.1.1 VCO Basics

An oscillator is a circuit that generates a periodic signal in its steady state. The frequency of a voltage-controlled oscillator (VCO) is controlled by an external voltage source. It has no RF input signal, and it depends on the circuit noise to initiate a growing signal that settles to stable periodic signal in steady state. Oscillators may be used for frequency conversion in transceiver circuits and for clock generation in digital systems. An oscillator that generates a sine wave is a harmonic oscillator. A cross-coupled LC oscillator is widely used in communication systems. Compared to the resonator-less ring oscillator, it has superior phase noise performance but poorer quadrature accuracy when used to generate quadrature signals. As shown in Fig. 2.1, the cross-coupled LC VCO is composed of two parts; an active cross-coupled pair and a tunable resonator including the passive elements.

K. Khalaf et al., *Data Transmission at Millimeter Waves*,
Lecture Notes in Electrical Engineering 346, DOI 10.1007/978-3-662-46938-5_2

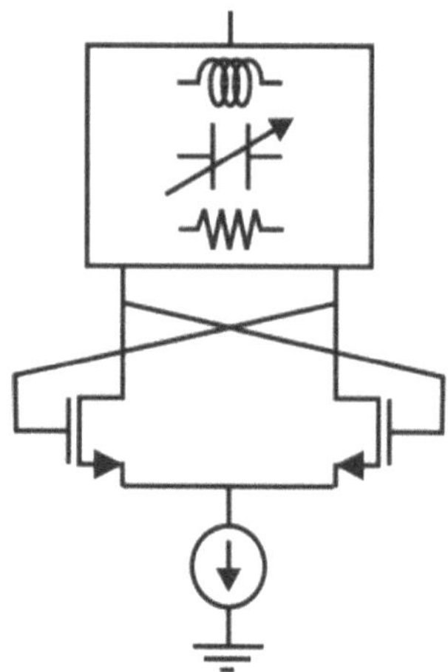

Fig. 2.1 Cross-coupled LC VCO

The transconductance and output resistance of the cross-coupled pair can be derived by connecting an AC voltage source at the output of the active part and deactivating independent sources, as in Fig. 2.2.

By relating V and I in Fig. 2.2, we can determine G_m and R_{out} as following:

$$G_m = \frac{I_{out}}{V_{in}} = \frac{I}{V} = -\frac{i_0}{2V_{gs}} = -\frac{g_m}{2} \tag{2.1}$$

$$R_{out} = V/I = -2/g_m = 1/G_m \tag{2.2}$$

where G_m is the total active transconductance, $g_m(=i_0/v_{gs})$ is the transistor transconductance and R_{out} is the total active output resistance.

The impedance seen at the drain terminals of the cross-coupled pair can now be seen as a negative resistance $-R_m$ (R_m is assumed to be a positive number) with a noisy current source, as shown in Fig. 2.3. The equivalent impedance of the tank circuit at resonance reduces to a resistor, because both positive (inductive) and negative (capacitive) reactances cancel each other.

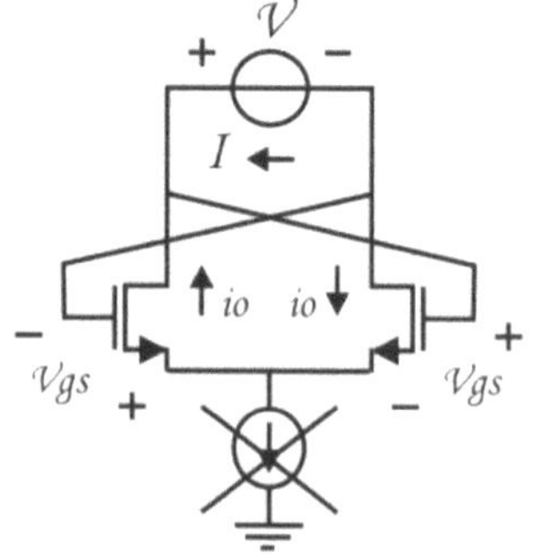

Fig. 2.2 Small-signal analysis of the active part

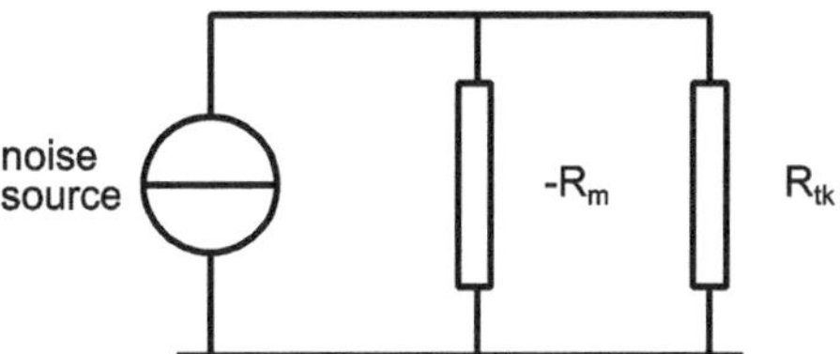

Fig. 2.3 Oscillator negative resistor model

The circuit will start oscillation, with the help of the noise source, when the negative resistance (resembling a power generating element) is higher in magnitude than the positive resistance (which is a power dissipative element) that represents tank losses. This is the oscillation condition, which is equivalent to saying:

$$G_m > \frac{1}{R_{tk}} \tag{2.3}$$

where R_{tk} is the tank resistance. This is a single port model for the oscillator. The factor by which the negative resistance is higher than the positive one (R_m/R_{tk}) is the oscillation margin. This value should be greater than unity to ensure starting of oscillation.

2.1.2 Main Parameters

The voltage-controlled oscillator is characterized by four main parameters: center frequency, tuning range, output voltage swing and phase noise. Figure 2.4 is a graphical illustration of these parameters.

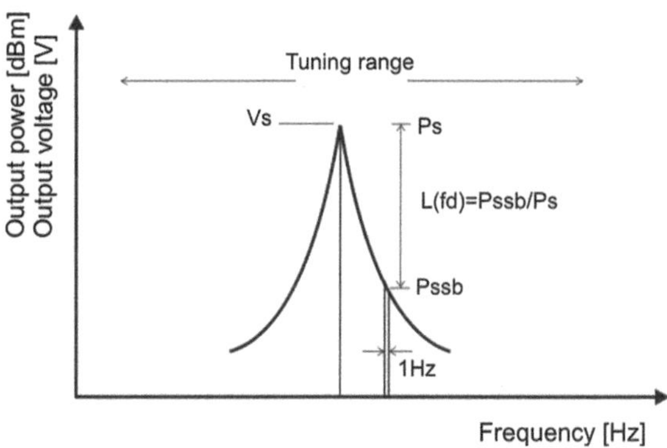

Fig. 2.4 VCO main defining parameters

The oscillator center frequency (f_0) is the frequency at which the output power is largest. This is defined by the resonant frequency of the tank circuit, which leads to the following result:

$$f_0 = \frac{1}{2\pi\sqrt{LC}} \tag{2.4}$$

where L and C are the total inductance and capacitance seen at the drain nodes of the cross-coupled pair.

The oscillator tuning range is the difference between the maximum and minimum output frequency of the oscillator ($f_{\max} - f_{\min}$). This is usually controlled by a varactor, where the maximum and minimum capacitance of the varactor corresponds to the minimum and maximum output frequency of the oscillator, respectively.

$$f_{\max} = \frac{1}{2\pi\sqrt{LC_{\min}}} \tag{2.5}$$

$$f_{\min} = \frac{1}{2\pi\sqrt{LC_{\max}}} \tag{2.6}$$

$$f_0 = \frac{f_{\max} + f_{\min}}{2} \tag{2.7}$$

The oscillator's output voltage swing is the amplitude at the oscillator center frequency. It should be high enough to drive the following stage. This is usually not the case, and so a buffer is needed to deliver the required amplitude to the load. This will be shown in detail Sect. 2.2. Voltage swing is usually limited by tank losses (R_{tk}), and can be calculated, assuming a square wave output current, using the following equation:

$$V_{s-peak-diff} = A \approx \frac{2}{\pi} R_{tk} I_{tail} \tag{2.8}$$

The main assumption to the previous equation is that the output current arises from ideal on-off switching of the transistors, and therefore the tail current is commutated between either sides of the oscillator. The current through R_{tk} is then as shown in Fig. 2.5. The tank circuit is tuned to the fundamental tone of the square wave, which is then multiplied by the tank resistance to give a sinusoidal, differential output voltage swing.

Equation 2.8 is only valid as long as the bias current is not large enough to push the tail transistor into the triode region. When the bias current is increased, the gate-source voltage of the tail transistor is increased while the drain-source voltage is limited by the voltage headroom available from the supply. At the edge of the triode region, the voltage swing is limited by the supply, and no longer proportional to the

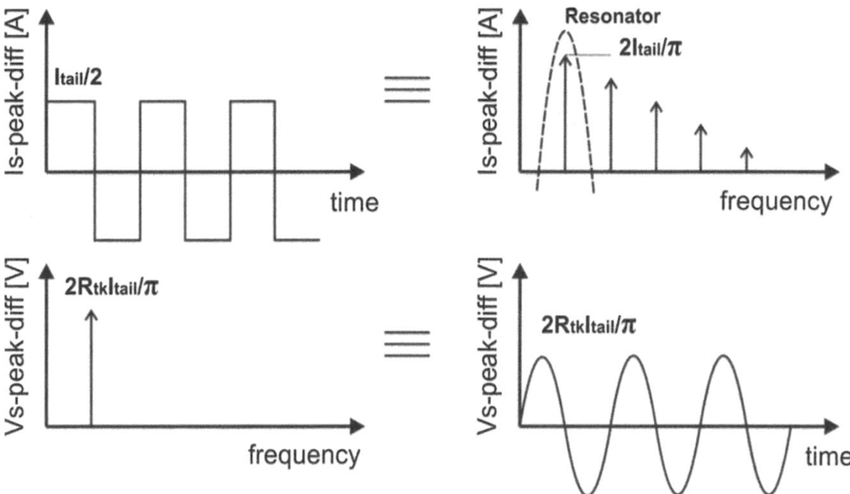

Fig. 2.5 Conversion from square wave current to sinusoidal output voltage through filtering by the resonator

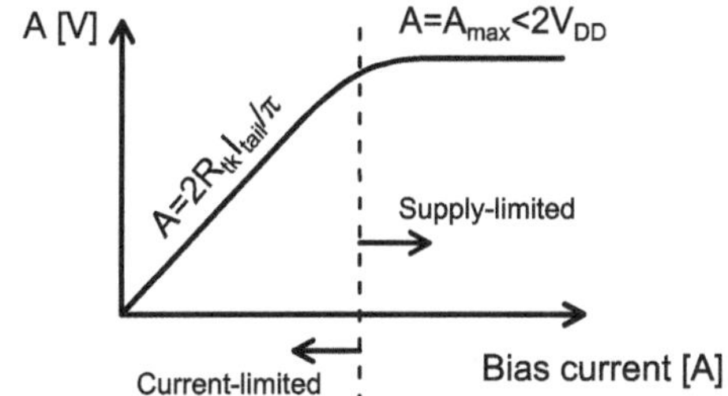

Fig. 2.6 Oscillator output differential amplitude based on the operation of the tail current transistor

tail current. Thus, two regions of operation are defined: the current-limited regime and the supply-limited regime [1]. The oscillator output differential amplitude within the two regions is shown in Fig. 2.6.

The spectrum of the output voltage signal of a real VCO circuit is not just a single frequency representing a pure sine wave. As shown in Fig. 2.4, the signal is spread in frequency having a skirt shape. In time domain, this can be seen as random variation of zero-crossings of the periodic sine wave signal representing the fundamental tone. In a phasor representation, this can be seen as a split into amplitude-modulated (AM) wave and phase-modulated (PM) wave as shown in Fig. 2.7, where both can yield phase noise, either directly or indirectly.

Phase noise is the parameter defining the spectral purity of the oscillator. The oscillator output signal is more "pure" when the fundamental component of its

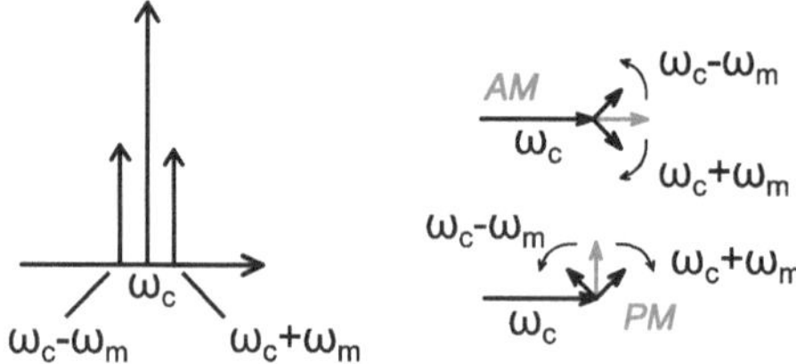

Fig. 2.7 Sidebands can be seen as AM and/or PM signals [2]

frequency domain is less spread in frequency. This is translated to a lower phase noise value. This parameter is very crucial, especially in receiver circuits. As shown in Fig. 2.8, an oscillator with a high phase noise can cause frequency down-conversion for unwanted adjacent channels, which cannot be distinguished from the wanted signal. This leads to signal interference and higher noise, reducing the system's signal-to-noise ratio (SNR).

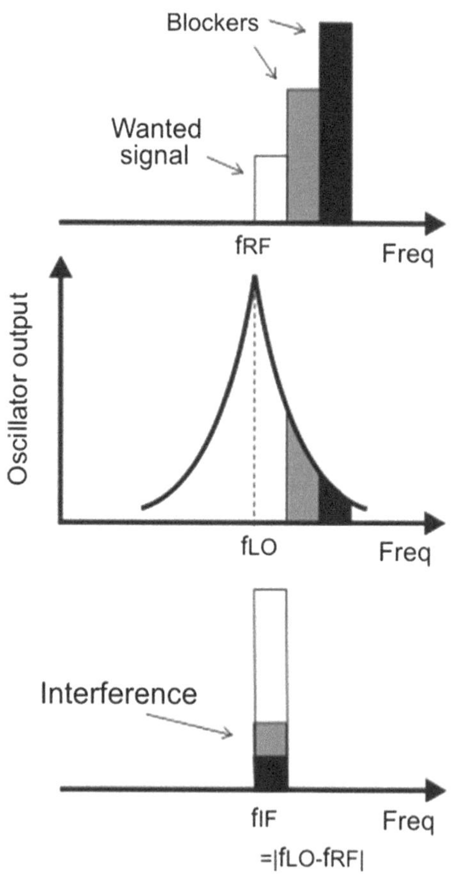

Fig. 2.8 Down conversion of unwanted frequency bands due to oscillator spectral impurity

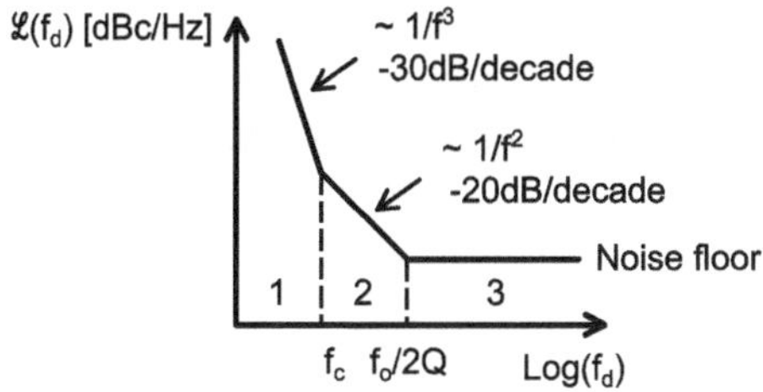

Fig. 2.9 Phase noise spectrum in dBc/Hz

As mentioned in [2, 3], phase noise is characterized using the single-sideband (SSB) phase perturbation spectral power in a 1 Hz bandwidth (spectral power density) at a frequency offset f_d away from the carrier frequency normalized to the power of this fundamental carrier frequency. Figure 2.9 shows the phase noise curve in dBc/Hz versus frequency. Three regions are defined according to the phase noise slope. The first region is independent of frequency, which is the white noise in the system. The second region is proportional to f^2, and it shows the tank effect on the thermally induced noise sources in circuit components. Close-in phase noise is represented by the third region, which is proportional to f^3 and is due to active elements' flicker noise up-conversion close to the carrier frequency.

Several models and analyses for phase noise are presented in the literature aiming at understanding the relationships between circuit parameters and phase noise, and getting an equation that can predict the phase noise value [4–8]. More intuitive interpretations and closed form formulas were also developed [9, 10]. This is in order to have the possibility of exploring ways to reduce this unwanted effect in oscillators. Phase noise in the $1/f^2$ region, assuming a square wave output current and neglected parasitic capacitance, can be written in terms of circuit parameters as following [11].

$$\mathcal{L}(f_d) = 10\log\left[\frac{kT}{C^2A^2R_{tk}f_d^2}(1+\gamma_n)\right] \tag{2.9}$$

where C is the total VCO capacitance at the output nodes, A is the peak differential voltage swing, R_{tk} is the equivalent losses at the oscillator output and γ_n is the NMOS transistor excess channel noise parameter. This equation accounts for both tank and switch noise. Accurate prediction of the phase noise value is not expected using this equation at 60 GHz due to large parasitics. However, it is useful in determining the effect of circuit parameters on phase noise in the $1/f^2$ region.

2.1.3 Phase Noise Origins

Understanding phase noise origins can help choosing the correct modification in a circuit to reduce its value. A brief summary of phase noise origins will be presented in this section. In [2] the differential LC oscillator phase noise is studied in great detail. The thermally induced phase noise can be a result of three main sources: the resonator, the differential pair and the current source.

2.1.3.1 Resonator Noise

Resistance R_{tk} representing tank losses is the noise generating element in the resonator. The noise current can be divided, due to the cross-coupled pair non-linearity, into AM and PM signals modulating the main oscillator tone. The AM signal can be filtered due to the limiting action of the cross-coupled pair. The cross-coupled pair negative resistance cancels the tank losses, and the PM signal is multiplied by the lossless resonator transfer function. This shows the importance of a lower bandwidth, i.e., higher quality factor resonator.

2.1.3.2 Differential Pair Noise

Noise in the cross-coupled pair will only be effective when both transistors are in the active region (this is mostly the case at 60 GHz). If one transistor switches off, the noise current of the other transistor will be in series with a constant current source I_{tail}, and thus eliminated. This can be modeled as the cross-coupled transistor white noise current multiplied by a pulsed G_m function with a frequency of $2f_o$. As shown in Fig. 2.10, the current noise of the cross-coupled transistors only at the fundamental frequency and its odd harmonics will cause noise to be folded at the oscillation frequency when multiplied by the pulsed G_m function. This analysis shows the importance of the noise generated at odd harmonics of the oscillation frequency from the cross-coupled pair. Note that the width of the time window at which both transistors are active doesn't affect the output referred noise density. The higher transistor transconductance, the less MOS transistors are in saturation region, and thus, the higher G_m sinc function bandwidth. But input-referred noise noise density is also lowered with higher transonductance. This leads to the same integrated rms output noise [12].

2.1.3.3 Tail Current Noise

Noise in the tail current will be commutated between the two sides of the oscillator. This can be modeled as being multiplied by a square wave with frequency components at the fundamental and odd harmonics.

Multiplications that will end up with noise components around the fundamental frequency are the square wave fundamental component with tail noise at DC and at second harmonic. Also the square wave third harmonic with the tail noise second harmonic, and so on. This is shown in Fig. 2.11. Note that only tail noise components at DC and even harmonics are causing noise components at the

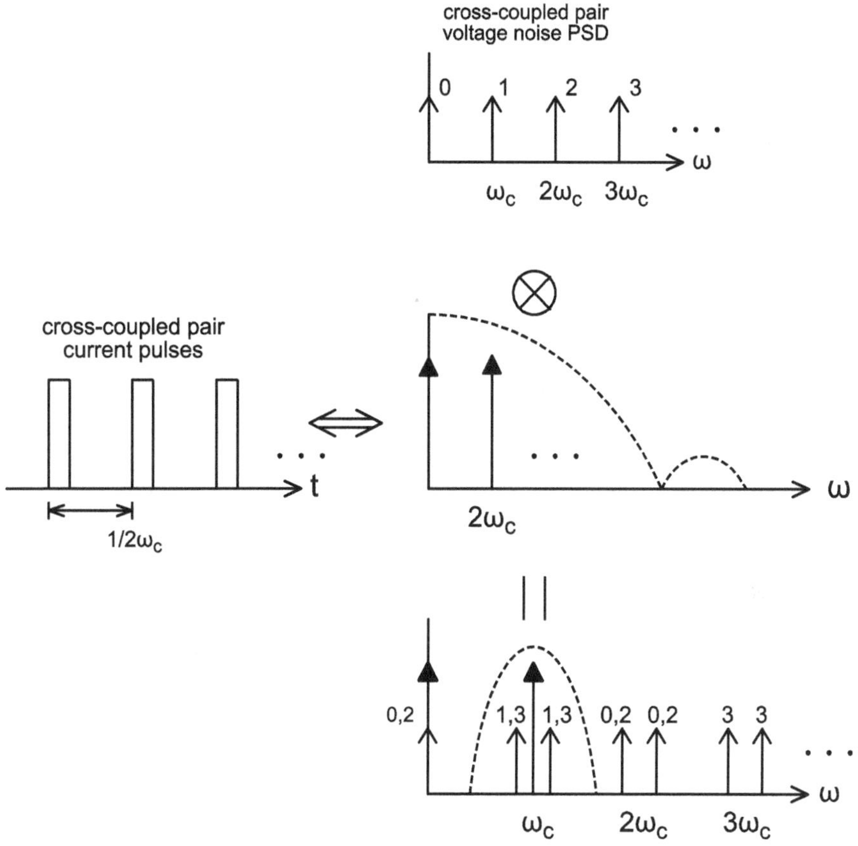

Fig. 2.10 Noise folding due to cross-coupled pair [2]

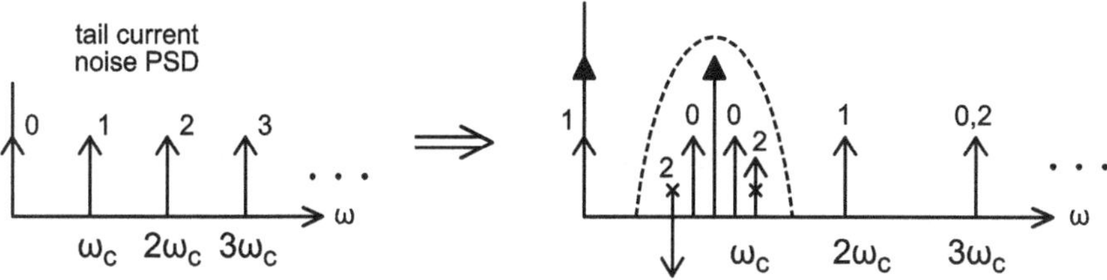

Fig. 2.11 Tail noise mixing with the cross-coupled pair [2]

fundamental frequency. Note also that tail noise component at DC will produce an AM signal. A varactor is a component that will convert voltage signal into a change in the capacitance value, and thus, a change in the operating frequency. Owing to the varactor in the VCO, the AM signal generated from the DC component of the tail noise can be converted into an FM signal which appears as phase noise around the center frequency [13, 14].

2.1.4 Quadrature VCO

In direct conversion receivers, positive and negative sidebands of the RF signal spectrum are down-converted on top of each other at baseband [15]. In frequency and phase modulated signals, two down-converting paths with a 90° phase shifted oscillator signal are needed for demodulation. Quadrature voltage-controlled oscillator (QVCO) uses coupling mechanisms between two VCO's in order to produce four-phase outputs, all orthogonal to each other (Fig. 2.12).

One more parameter can be defined for the QVCO beside the main VCO parameters described before in Sect. 2.1.2: phase error or quadrature error. For multi-phase oscillators, phase error is the difference in degrees between the actual phase difference between two subsequent output terminals in the oscillator and the ideal value. In the QVCO with in-phase (I) and quadrature (Q) outputs, quadrature error is the deviation from the 90° phase difference between I and Q terminals.

Cross-coupled LC VCO's can be coupled in three different ways, each with its pros and cons: parallel coupling (P-QVCO), series coupling (with two different choices; TS-QVCO and BS-QVCO for top and bottom, respectively) and gate-modulated coupling (GM-QVCO) [16]. All of the main VCO parameters are affected by inserting the coupling transistor in the VCO core.

In the P-QVCO shown in Fig. 2.13, the coupling transistors are connected in parallel to the cross-coupled transistors. A, B, C and D outputs represent phase shifts of 0°, 180°, 90° and 270°, respectively. This topology is simple but has some disadvantages: phase noise is relatively high compared to the other topologies, and there is a trade-off between the phase noise and the quadrature accuracy through the coupling strength. The greater the coupling coefficient, the higher the phase noise but the better phase error, and vice versa.

The P-QVCO phase noise can be improved by independently biasing the gate of the cross-coupled pair [17]. This requires gate decoupling capacitors and biasing resistors as shown in Fig. 2.14. With a reduced gate voltage, the cross-coupled pair is allowed to provide more output voltage swings while operating in the saturation region.

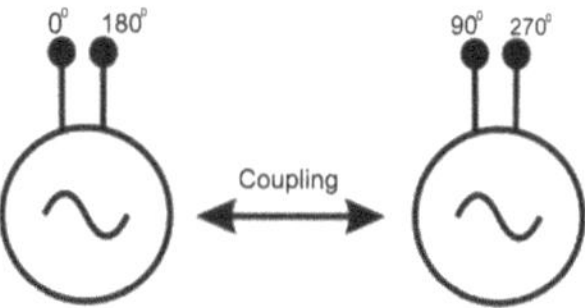

Fig. 2.12 Orthogonal signal out of the QVCO

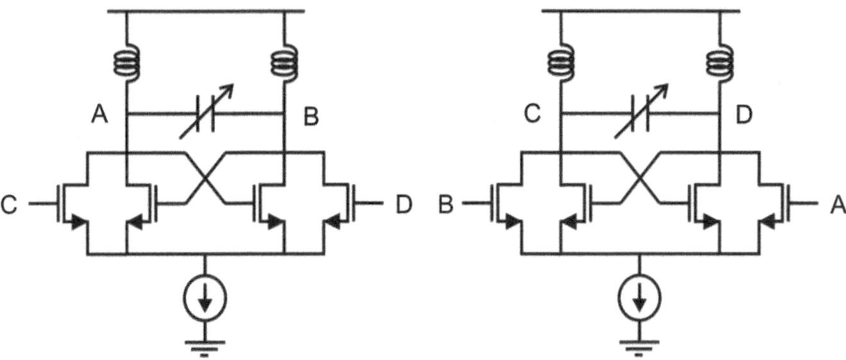

Fig. 2.13 Parallel QVCO (P-QVCO) topology

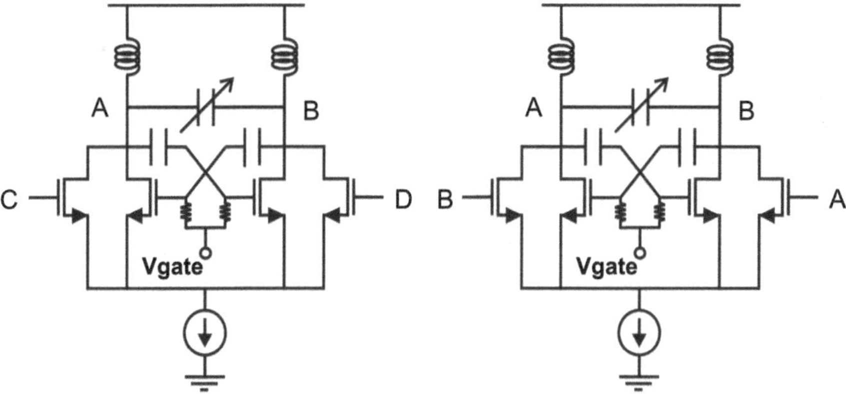

Fig. 2.14 P-QVCO with gate decoupling and external bias

In the top and bottom series-QVCOs of Fig. 2.15, coupling transistors are inserted in series with the cross-coupled pair. This takes from the voltage headroom available which is not so suitable for low-voltage applications. In the TS-VCO, large coupling transistors are needed to have lower phase error, which will dramatically increase the parasitic capacitance, and thus, reduce the tuning range. The BS-VCO, on the other hand, has higher phase accuracy and lower phase noise compared to the top-stacked one.

As shown in Fig. 2.16, a gate-modulated QVCO topology was proposed in [16]. The coupling transistors are placed in series with the gates of the switching transistors. This will improve the voltage headroom as compared to the series topologies. The GM-VCO was claimed to have the best quadrature accuracy and phase noise performances over the parallel and series ones. To ensure enough coupling strength from the opposite oscillator core, the coupling transistor sizes may need to be increased, which will lead to higher output parasitic capacitance, and thus, reduced tuning range.

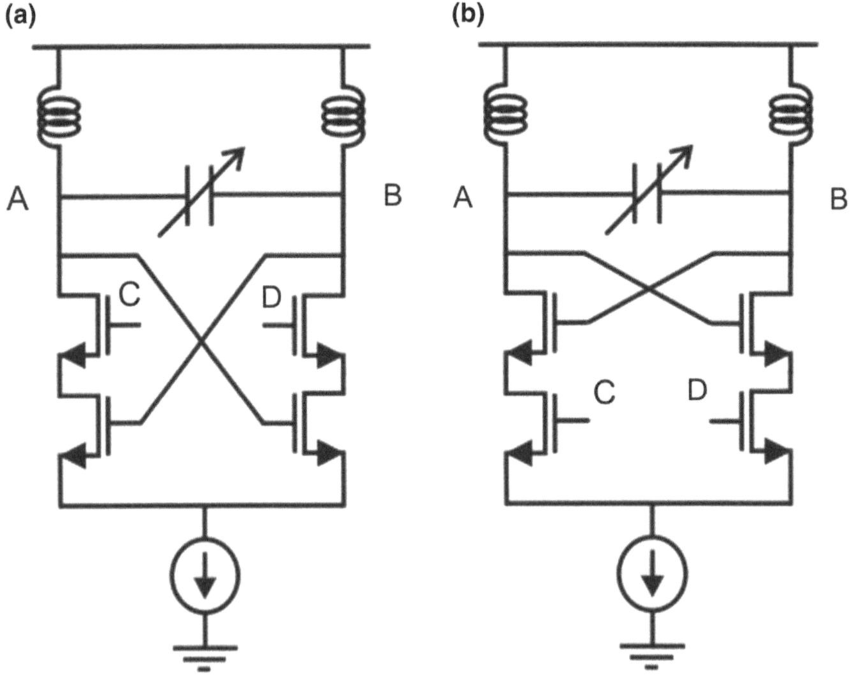

Fig. 2.15 Half-sections of series-QVCO in a top (TS-QVCO) and b bottom (BS-QVCO) configurations

2.2 LO Buffer

A buffer is usually needed after the VCO to minimize any effect of the output load on the oscillator signal. The VCO output can either feed another block in the system or go directly to the output for measuring purposes. In both cases, the VCO load can be modeled as a parallel combination of a capacitance and a resistance. The load capacitance can reduce the oscillation frequency and tuning range. The load resistance, however, can reduce the output amplitude. Thus, the phase noise can also be increased. Buffers are also needed to increase the output amplitude. Local oscillator (LO) buffers, for example, can deliver the output signal to a mixer. For higher conversion gain, the mixer input amplitude should be increased. LO buffers can be useless if it has a higher load than the following stage or if the VCO amplitude is large enough for the operation of the following circuit.

Two transistor configurations can be used as buffers for the VCO: source followers and common-source (CS) amplifiers. Source-followers reduce the VCO output amplitude (Fig. 2.17). They can be used if the VCO output is going to be directly measured stand-alone. In this case, large output swing is not required as it is used for testing purposes. The common drain transistor has a low output resistance, which is suitable for driving the output 50 Ω load.

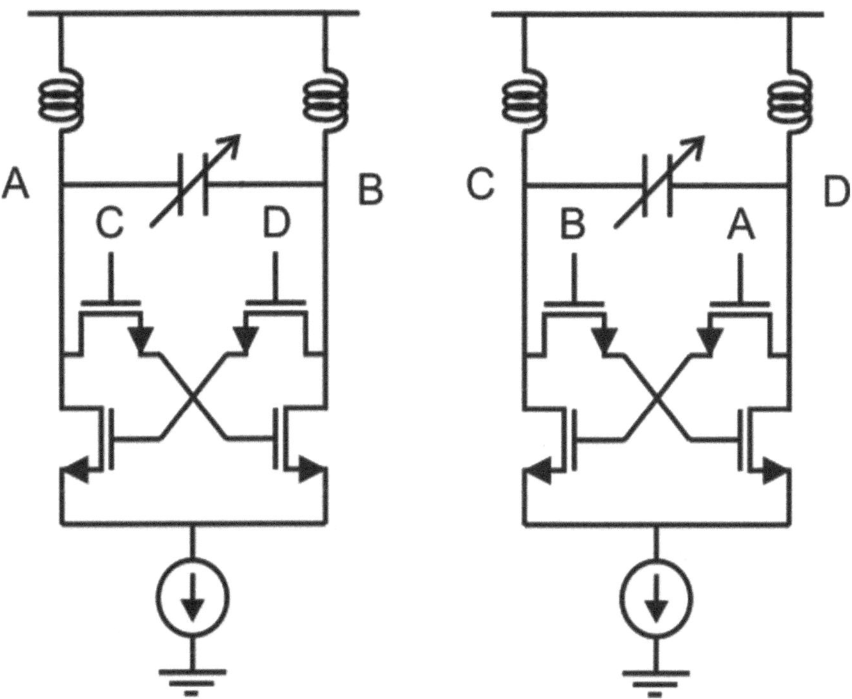

Fig. 2.16 Gate-modulated QVCO (GM-QVCO) architecture

Fig. 2.17 Source follower buffer

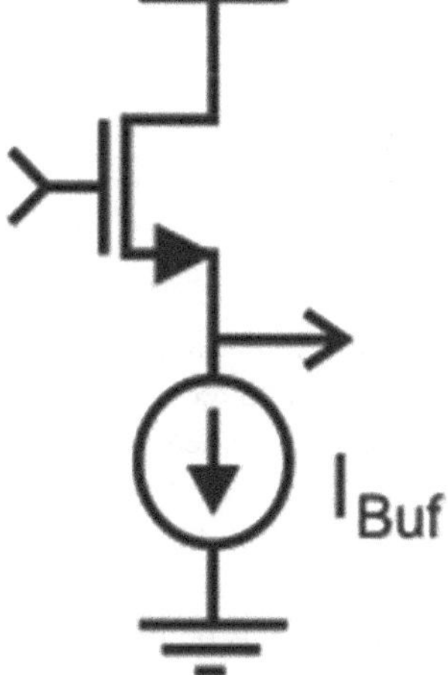

Common-source amplifiers can also be used after the VCO for buffering, as shown in its differential form in Fig. 2.18. Inductors can be used at high frequency to tune out all the parasitic and load capacitances at the output node. This allows the transistor to

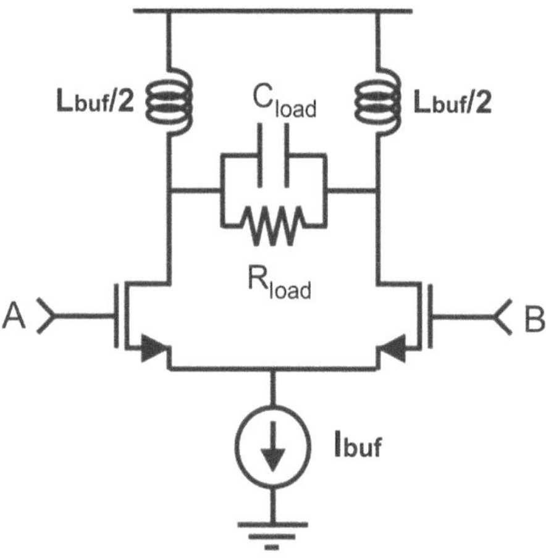

Fig. 2.18 Inductively-tuned CS differential amplifier as a buffer

amplify the input signal within the required frequency range, with a peak at the tuning frequency f_{tune}, and a bandwidth limited by the current source I_{buf}.

$$f_{tune} = \frac{1}{2\pi\sqrt{L_{buf}C_{tot}}} \tag{2.10}$$

where $C_{tot} = C_{parasitic} + C_{load}$. When the buffer is used for measuring, an accurate prediction of the pad capacitance is required for choosing the buffer inductor value. Any mismatch between the buffer tuning frequency and the oscillator frequency will cause a significant reduction in the output amplitude.

The small-signal model of the buffer is shown in Fig. 2.19. If the inductor cancels all capacitive elements at the output node, the buffer gain can be calculated as following:

$$A_{buf} = G_m(r_{out}||R_{load}||R_{par,L}) = \frac{g_m}{2}R_{out} \tag{2.11}$$

If the total capacitance at the output node is not large enough, large inductor values will be required. Maximum inductance is usually limited by the inductor self resonance frequency, after which the lines forming the inductor behave capacitively. One way to get the gain peak at the required frequency is to add more capacitance at the output. Any additional capacitance comes with its parasitic resistance. This will add more load resistance to the output, and the total parallel resistance will be reduced, causing gain reduction.

$$A_{buf} = G_m(r_{out}||R_{load}||R_{par,L}||R_{par,C}) \tag{2.12}$$

Another way to get the tuning frequency with a limited inductor is to exchange the inductor load with a transformer as in Fig. 2.20a. The transformer used is

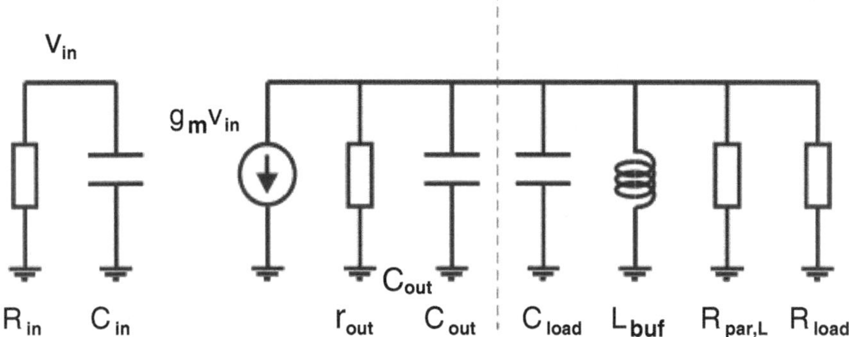

Fig. 2.19 Model of the active buffer

nothing but an increased equivalent inductance with the factor (k), which is the coupling coefficient. So:

$$L_{eq} = L_{tune}(1 + k) \tag{2.13}$$

It is worth noticing that a transformer is usually implemented with a lower quality factor than the inductor, as more than one metal layer should be used compared to the only top metal layer used in the inductor implementation. The gain can be the same as in Eq. 2.11 with a different inductor quality factor.

$$A_{buf} = G_m(r_{out}||R_{load}||R_{par,trafo}) \tag{2.14}$$

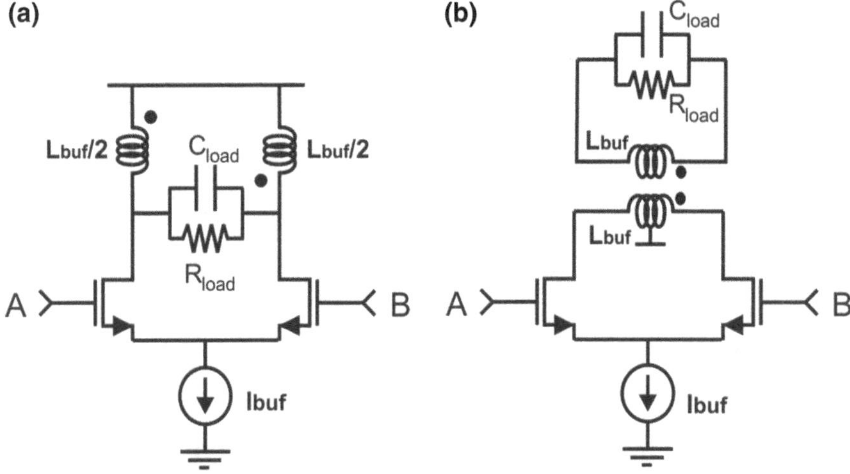

Fig. 2.20 Active buffer with transformer load **a** voltage output **b** current output

As shown in Fig. 2.20b, the transformer can be used in such a way that the output current of the common-source transistor is used instead of the output voltage. One side of the transformer will be connected to the buffer circuit, and the other side will be connected to the load.

This transformer-coupled differential amplifier is analyzed in [18]. If the load is assumed to be only capacitive, it will be transformed to the buffer output node with an equivalent impedance value that is elaborated in Appendix A, and given below in its final form:

$$Z_{out} = j\omega\left(L + \frac{\omega^2 L^2 C_{load} k^2}{1 - \omega^2 L C_{load}}\right) \tag{2.15}$$

This means that for practical values (for example, f = 60 GHz, L = 100 pH and $C_{load} = 20\,fF$), the denominator will always be positive, and the common-source transistors will see an equivalent inductance value that depends on the load. Note that the equivalent inductance is higher than the primary value of the transformer. In practice, the buffer load is a transistor with an equivalent input parallel capacitance and resistance. The resistive component is transformed to the buffer output with a higher value (R'_{load}) [18]. Thus, the buffer voltage gain can be calculated as following:

$$A_{buf} = G_m(r_{out}||R'_{load}||R_{par,trafo}) \tag{2.16}$$

Note that the voltage-output transformer-coupled buffer is expected to provide higher gain (Eq. 2.14) compared to the current-output one (Eq. 2.16) because of the higher load resistance.

2.3 Frequency Divider

Frequency dividers are circuit blocks used to divide an input signal in the frequency domain. They can be categorized into static and dynamic dividers. Static dividers use bi-stable latches and, for operation at high frequencies, can be implemented using current-mode logic (CML) circuits [19]. Dynamic dividers don't quantize the divided signal in either amplitude or time. They are divided into regenerative, parametric and harmonic injection dividers [20]. The harmonic injection dividers are of interest because they can operate at smaller input signal amplitudes [20]. They depend on a free-running oscillator, and synchronizing the harmonics of the free-running frequency with an input source.

Static dividers have a trade-off between speed (and thus maximum input frequency) and power dissipation, and they can operate down to DC. Analog dividers, on the other hand, can operate at higher input frequencies with lower power consumption using only few transistors, but usually with limited input bandwidth (locking range).

2.3.1 ILFD

Oscillators depend on the non-linear behavior of circuit components to reach their steady-state. This non-linearity will enable harmonic components to appear together with the fundamental oscillation frequency. An input source can be injected at any of these harmonic frequencies, and synchronization of the oscillator output (i.e., injection locking) can take place. Locking range will decrease with higher order of the oscillator harmonic components because they have lower amplitudes.

Harmonic injection dividers are one group of injection-locked oscillators (ILOs). ILOs are divided into three categories; first-harmonic, sub-harmonic and super-harmonic ILOs. This depends on the relationship between the input signal frequency and the free-running oscillator frequency. The input frequency is the same as the oscillator free-running frequency in the first-harmonic ILO, lower and higher in the sub-harmonic and super-harmonic ILOs, respectively. So, harmonic injection dividers are super-harmonic ILOs, and they're also called injection-locked frequency dividers (ILFDs).

ILFDs can be modeled as shown in Fig. 2.21 [21]. The model includes a non-linear device that generates harmonic energy and a band-pass filter (BPF) to select one of these harmonics. The BPF output is then fed back to the non-linear device and oscillation keeps running independently. An input signal can then be injected in the oscillator signal path to be synchronized with the selected frequency component after the BPF.

As the input signal frequency changes, the output should follow this change. The range of input frequencies across which the oscillator is still locking and the signal is divided correctly is the locking range. A large locking range is important, as the frequency divider should cover the tuning range of the VCO plus a good margin. At high frequencies, larger margin is required to ensure proper operation within process, voltage and temperature (PVT) variations in the circuit.

ILFDs can be implemented using a cross-coupled LC oscillator generating the free-running signal. Traditional ILFDs [21] inject the input signal at the gate of the tail current transistor as shown in Fig. 2.22. They suffer from large input capacitance, small locking range and they operate at low input frequencies. This is due to the large tail transistor size. A shunt peaking inductor and capacitor were inserted at the common-source node of the cross-coupled pair to tune out the tail transistor output capacitance [22]. This solution improved the maximum frequency and locking range, but with the use of large area passives and the need for careful

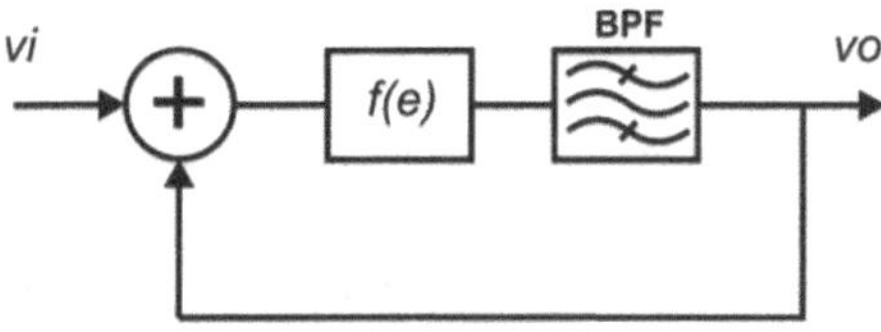

Fig. 2.21 Harmonic injection (injection-locked) frequency divider model [21]

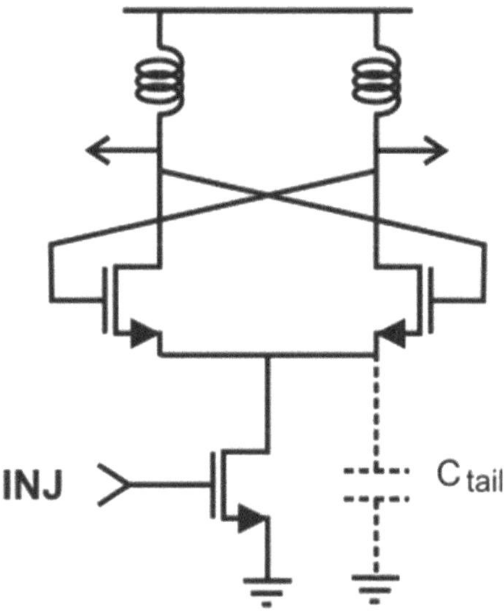

Fig. 2.22 Conventional ILFD

adjustments of the inductor and capacitor values to get the required parasitic cancellation.

Another way to inject the input signal is through a transistor switch connected in parallel to the tank as shown in Fig. 2.23a [23–25]. The direct ILFD doesn't incorporate extra passives and provides a simpler circuit. The injecting signal modulates the oscillator output and the signal with frequency difference is selected by the tank. A block diagram explaining the behavior of the circuit is shown in Fig. 2.23b [26]. The output signal (fi/2) is fed back and mixed with the input signal (fi) generating the sum (3fi/2) and difference (fi/2) of both signal frequencies. The band-pass filter selects fi/2 and passes it to the output, thus providing division. The transistor switch in Fig. 2.23a works as a drain-pumped mixer [27], and the cross-coupled pair with the tuning inductor form the feedback loop.

An analytical model for the direct ILFD is developed in [28]. The model depends on substituting the switching transistor (M_{in}) with passive elements. Figure 2.24 shows M_{in} and the relationship between the injected input voltage (V_{in}), the differential output voltage $V_{out}\pm$ and the channel current of M_{in} (I_{in}). The difference in phase shift between the input and output voltage signals is φ. The voltage and current waveforms for $\varphi = \pi/2$ and $\varphi = \pi/4$ are shown in Fig. 2.25. The locking range derived equation is as following:

$$\Delta\omega = 2g_{q,max}/C = 2\omega_0^2 L g_{q,max} \tag{2.17}$$

where L and C are the tank inductance and capacitance, respectively. $g_{q,max}$ is the equivalent injecting transistor output conductance ($g_{q,max} = I_q(\varphi)/2v_o$), which appeared as a result of modeling the injecting transistor as an inductor or a capacitor

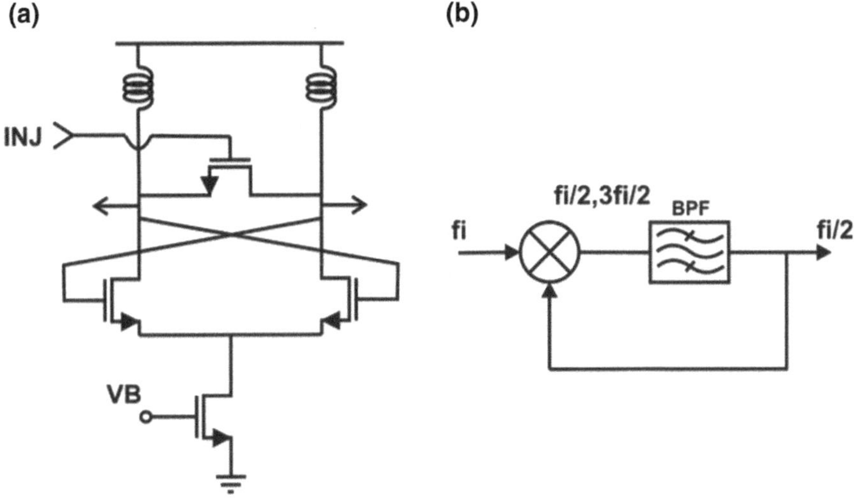

Fig. 2.23 Direct ILFD **a** circuit schematic and **b** equivalent model [26]

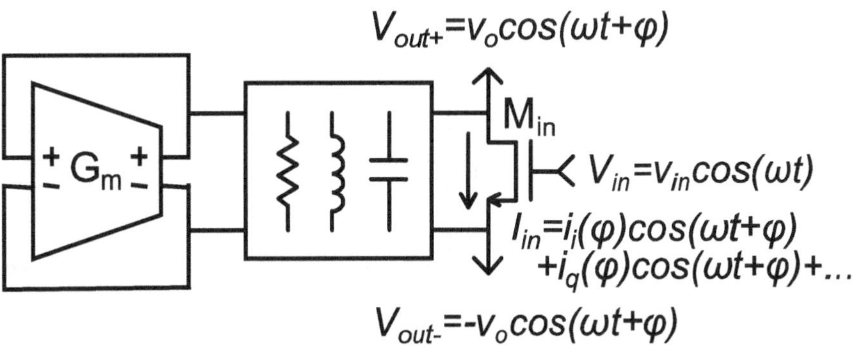

Fig. 2.24 Block diagram of the differential direct ILFD [28]

in parallel with a resistor. $I_q(\varphi)$ is the magnitude of the quadrature component of I_{in}, and v_o is the magnitude of the output voltage.

A Direct ILFD provides lower input capacitance and can operate at higher frequencies compared to the conventional one due to the smaller injecting transistor. Series peaking inductors were added in [29] to decrease the divider output capacitance and improve the locking range. Another approach that doesn't incorporate passive components is using two injecting transistors [26]. As shown in Fig. 2.26, the parasitic capacitance contribution of the injecting transistors to the divider output nodes is halved compared to using a single injecting transistor as in

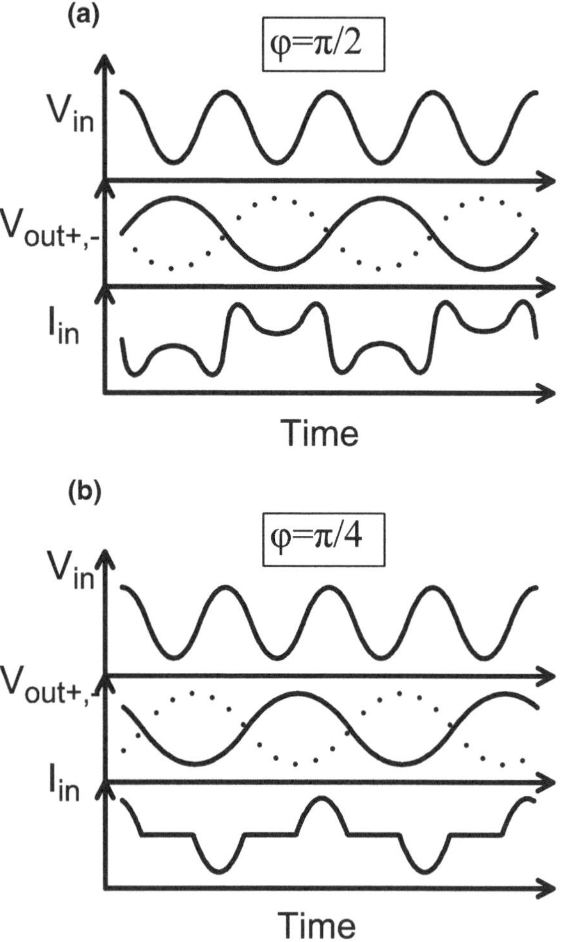

Fig. 2.25 Voltage and current waveforms **a** at $\varphi = \pi/2$ and **b** $\varphi = \pi/4$ [28]

Fig. 2.23a. This allows doubling the injecting transistor sizes at the same output parasitic capacitance. Thus, the dual mixing technique is used to double the effective injecting conductance.

2.3.2 Static Divider

Digital static dividers at high frequencies depend in their implementation on CML circuits. It consists of three main parts, pull-up load, pull-down network (PDN) and a current source [30]. The circuit behavior is described depending on the logic blocks in the PDN and the input combination. The basic element of the static divider is a D-flip-flop (DFF). The DFF inverted output can be connected to the

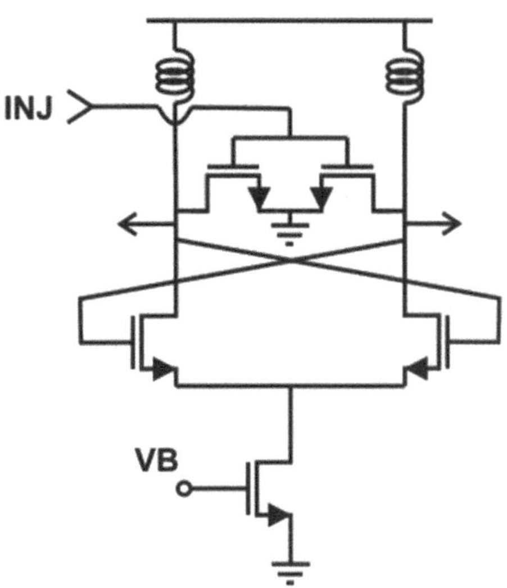

Fig. 2.26 Dual-mixing direct ILFD circuit schematic

input terminal and the input signal connected to the clock terminal to form a divide-by-two.

Two level sensitive latches in master-slave configuration can be used to form the DFF required for the division. As shown in Fig. 2.27, the first stage is a gated D latch [31] that is transparent through a differential pair buffering the input signal when the CLK signal is high. When the CLK signal is low, the circuit is

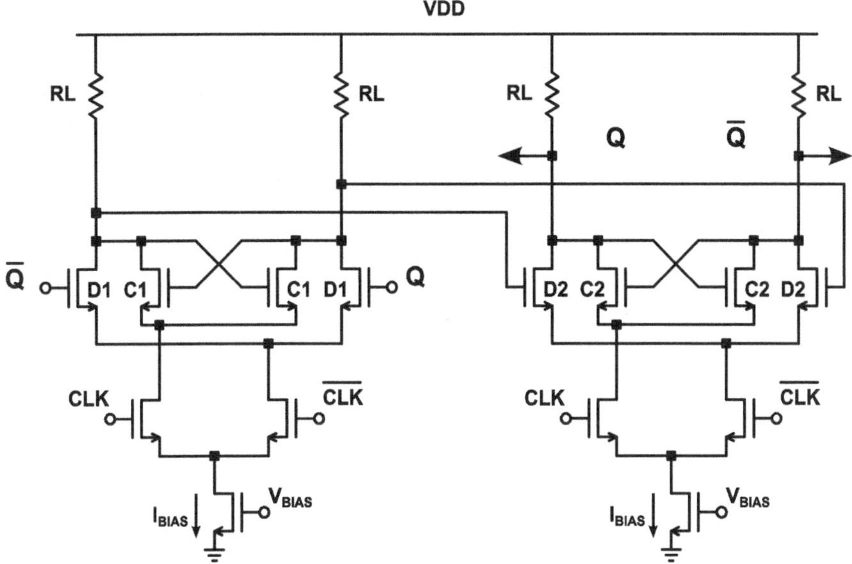

Fig. 2.27 Conventional CML latches in a master-slave configuration [32]

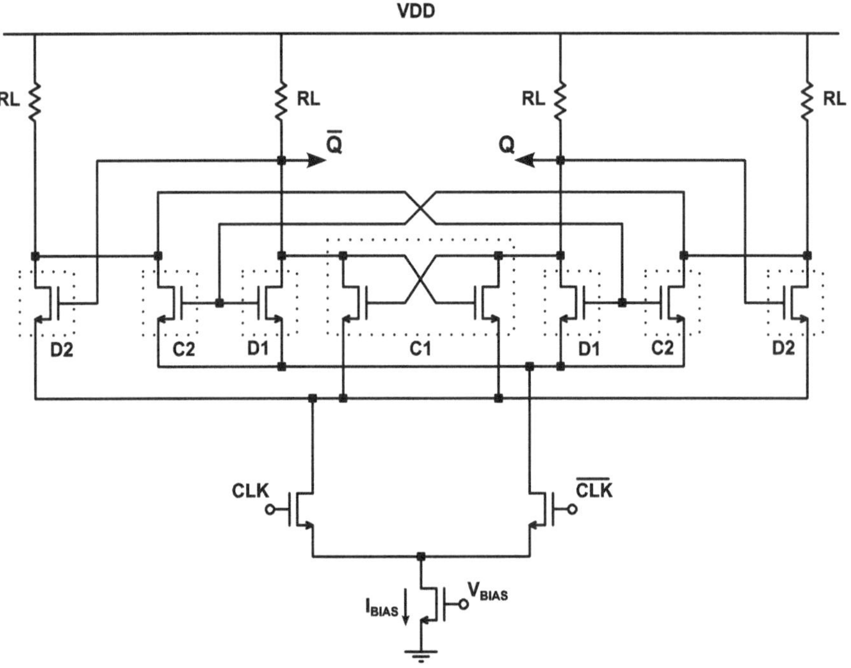

Fig. 2.28 High frequency CML divider (by two) [32]

non-transparent and the cross-coupled pair keeps the output state unchanged. The second stage works in the same way with inverted clock signals to implement the DFF.

The maximum operating frequency of the divider is limited by the CLK-Q time delay, which is a function of the total output capacitance and the load resistance, as well as the bias current. In [32], the cross-coupled pair size is reduced (to reduce the output capacitance) and the circuit is rearranged to have one tail transistor as shown in Fig. 2.28.

2.4 LNA

The low-noise amplifier (LNA) is usually used as the first block in the receiver front-end. It should add the lowest possible noise to the input signal. Noise degradation is usually measured with noise figure (NF). NF is a parameter that shows how much noise a block is adding to the system. Noise factor (F) is the linear equivalent of NF. LNAs should provide enough gain to overcome the noise figure of the following stages. This is suggested by Frii's formula, which calculates the system noise factor as following:

$$F_{total} = F_1 + \frac{F_2 - 1}{G_1} + \frac{F_3 - 1}{G_1 G_2} + \frac{F_4 - 1}{G_1 G_2 G_3} + \cdots \tag{2.18}$$

where G is the power gain of a block, and the subscript indicates the order of the block in the receiver. Assuming the LNA to be the first block in the system, Eq. 2.18 shows how the LNA (with noise facto F_1) is dominating the total noise, especially with a high gain (G_1) value.

The LNA input should be matched to 50 Ω to provide the lowest possible reflections from the source. It shouldn't also distort the input signal. Signal distortion is caused by the non-linear behavior of a block. Non-linearity is usually specified by the third order input-referred intercept point (IIP3). The total IIP3 of a system can be calculated as following:

$$\frac{1}{IIP3_{total}} = \frac{1}{IIP3_1} + \frac{G_1}{IIP3_2} + \frac{G_1 G_2}{IIP3_3} + \frac{G_1 G_2 G_3}{IIP3_4} + \cdots \tag{2.19}$$

Equation 2.19 shows that non-linearity of the latter stages are more effective due to the gain of the previous stages. So, the LNA distortion is not dominating the system non-linearity.

2.4.1 *NF and IIP3*

Noise figure of a linear two-port network as a function of the source admittance can be represented by:

$$F = F_{min} + \frac{R_n}{G_s} \left| Y_s - Y_{opt} \right|^2 \tag{2.20}$$

where F_{min} is the minimum achievable noise factor, Y_s (= G_s + jB_s) is the source admittance, Y_{opt} is the optimum load at which F reduces to Fmin (noise match condition) and R_n is the noise resistance defining the sensitivity of F to changes in the source admittance.

Note that these parameters can be related to circuit parameters, such as f_T, g_m and C_{gs} for a MOS transistor [33]. For minimum noise figure, F_{min} should be minimized by choosing the correct bias point, and the LNA input should be matched to the optimum source impedance that gives the minimum noise factor (Z_{opt}). The source impedance for noise match is usually not 50 Ω leading to the either a compromise between impedance and noise matching conditions or using a topology that allow for choosing the two impedances independently.

Non-linearity will cause additional tones to be generated at harmonic frequencies. If a signal with two frequency components at f_1 and f_2 enters the amplifier, more frequency components appear in the frequency band. Figure 2.29 shows the

Fig. 2.29 Two-tone excitation resulting tones (to the third-order) [34]

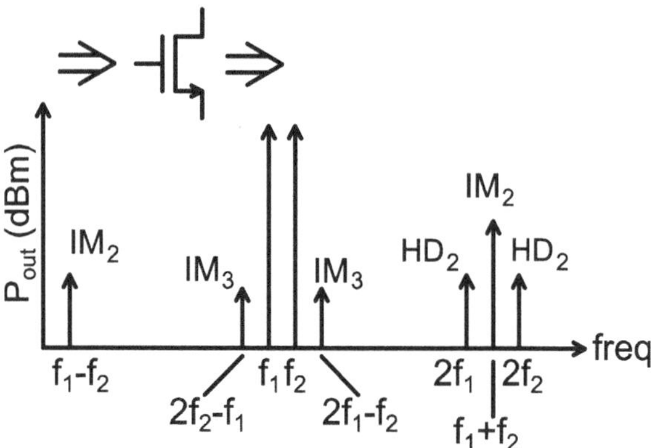

output spectrum with additional frequency components due to non-linearity (only to the third order).

Harmonic frequencies ($2f_1$, $2f_2$, $3f_1$, $3f_2$, …) and second-order intermodulation components ($f_1 - f_2$ and $f_1 + f_2$) are of less importance as they can be easily filtered out. In a direct conversion receiver, ($f_1 - f_2$) falls in-band but is usually not effective when using differential circuits. The third-order intermodulation products (IM3) are used in the definition of system non-linearity.

As shown in Fig. 2.30, the fundamental output tone eventually goes into compression with increasing input power. Linear extrapolation of the fundamental and IM3 curves will intersect at the third-order intercept point (IP3). Referred to its input, the IIP3 is used to define non-linearity in a system. The point at which the fundamental tone is compressed with 1 dB is the −1 dB compression point (P-1 dB), which is also used to define the non-linearity of a system. The P-1 dB is

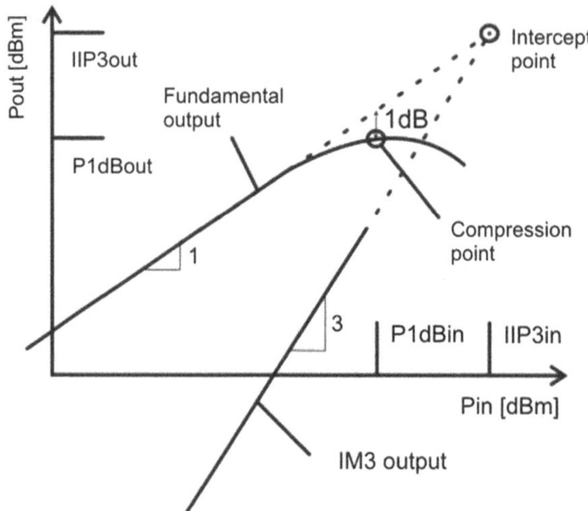

Fig. 2.30 Definition of important linearity parameters

easier to measure because it uses a single input tone, compared to the two-tone test for the IIP3 measurement. Input −1 dB compression point (P-1 dB,in) is around 10 dB lower than IIP3 [15], which gives an approximate value for the IIP3 when measured. Note that when dealing with a mixer, Fig. 2.30 is used with the x-axis (input power) at RF frequencies, while the y-axis (output power) is at the intermediate frequencies (IF) resulting after the frequency conversion.

2.4.2 LNA Topology

The commonly used topology for the LNA is based on a common-source transistor with inductive degeneration, as shown in Fig. 2.31a. If the small signal model of the transistor only contains an input capacitance C_{gs} and an output transconductance (Fig. 2.31b), the degenerated inductor can be transformed to the input using the β-transformation concept, leading to the following input impedance:

$$\begin{aligned} Z_{in} &= \frac{1}{j\omega C_{gs}} + j\omega L_{ss}[1 + \beta(\omega)] \\ &= \frac{1}{j\omega C_{gs}} + j\omega L_{ss} + j\omega L_{ss}\frac{\omega_T}{j\omega} \\ &= L_{ss}\frac{g_m}{C_{gs}} + j\left(\omega L_{ss} - \frac{1}{\omega C_{gs}}\right) \end{aligned} \tag{2.21}$$

where β(ω) is the current gain.

The input impedance contains a resistive part, which can be made equal to 50 Ω, and a reactive part. As the inductor L_{ss} is chosen to vary the resistive part, the reactive part will usually have a non-zero value. As the input capacitance C_{gs} is a very small value, the reactive part is usually capacitive. An inductor inserted at the

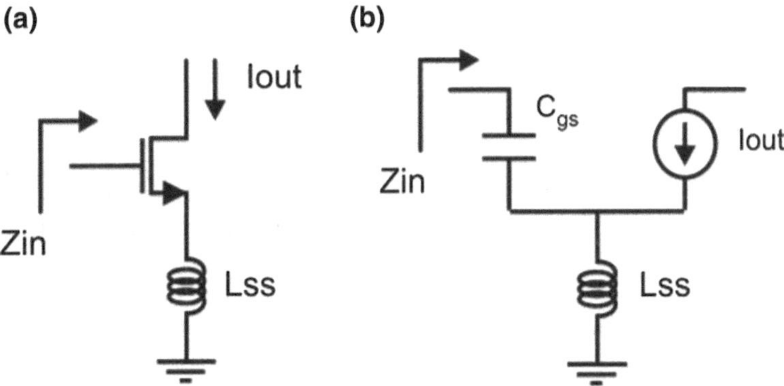

Fig. 2.31 **a** Inductively degenerated CS transistor and **b** small-signal model

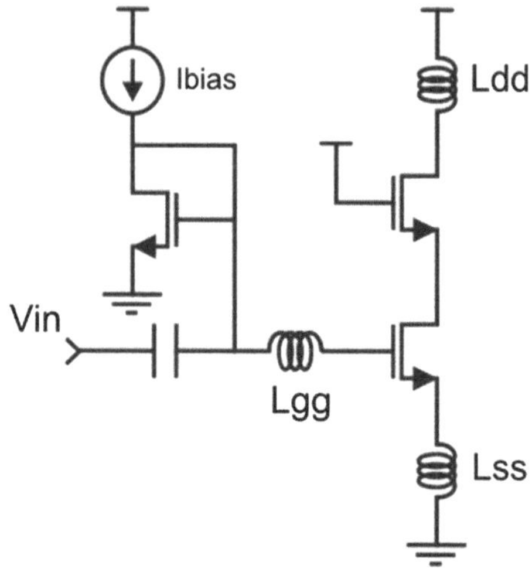

Fig. 2.32 Single-ended cascode LNA using inductive degeneration

gate can be used to cancel the imaginary part of the input impedance, leaving only 50 Ω to match the source impedance of the LNA.

In our small-signal analysis to get the LNA input impedance, we neglected a lot of components. When added to the small-signal model, the transistor output resistance, through the overlap capacitance, can cause a significant drop in the real part of the input impedance [33]. This is due to the path created to the load of the LNA. A cascode transistor (maybe with a larger gate length leading to a higher output resistance) can be used to isolate the output load from the input circuit. This can keep good input matching properties for the LNA with the drawback of additional noise figure. The complete LNA can now be as shown in Fig. 2.32.

LNA design can now be simplified to adjusting the transistor width for noise match while keeping minimum gate length for maximum gain. Then, we can adjust L_{ss} to have real input impedance equal to the source impedance. And finally, L_{gg} can be chosen to cancel the imaginary part of the input impedance. Thus, impedance and noise matching can "ideally" be achieved.

2.5 Mixer

After the received signal is amplified by a low-noise block, a down-conversion mixer is then used to bring the RF signal down to low frequencies. Signal processing at baseband is much easier and economical from the chip area and power consumption point of views. So, the LO generated signal is multiplied by the low-noise amplified RF signal via the mixer, and the signal with frequency difference is filtered at baseband.

2.5.1 Main Parameters

As Eq. 2.19 suggests, the receiver blocks closer to baseband have more effect upon the total linearity. Thus, mixer distortion usually dominates the system non-linearity.

Two noise figure definitions are common in the mixer: single-sideband (SSB) and double-sideband (DSB) noise figures. In non-zero IF systems, the input frequency band includes the required RF signal and maybe another signal at the same distance from the LO signal as that between LO and RF signals. This is called the image frequency. Both frequencies, the RF and image, can down-convert to the lower IF frequency band, because they're at equal distance from the LO signal (on opposite sides). Noise from both frequency bands down-convert to the same frequency and contribute to the output noise. If we assume a noiseless mixer and useful information exists in the image band as well as the RF band, then the noise factor is $SNR_{in}/SNR_{out} = (Pi \times No)/(Po \times Ni) = 1$ (NF = 0 dB). This is the way how DSB NF is calculated. In the SSB NF calculation, it is assumed that the image band doesn't include useful information (which is the usual case). So, the SNR at the output is doubled, because there is only noise coming from the image band. This will cause the noise factor to be 2 (NF = 3 dB). The two situations can be graphically illustrated as in Fig. 2.33. Unless otherwise specified, the DSB NF is usually used to define the noise figure of the mixer.

The input and output signals of the mixer are not at the same frequency. Thus, the conversion gain (CG) parameter is used in the mixer if it is providing gain (otherwise, conversion loss). CG is defined as the ratio of the desired IF output to the value of the RF input at a given LO signal level [3]. CG can be defined in the voltage domain (CG_v) or power domain (CG_p), and they're related through the ratio of the RF and IF port impedances, as shown in the following equations:

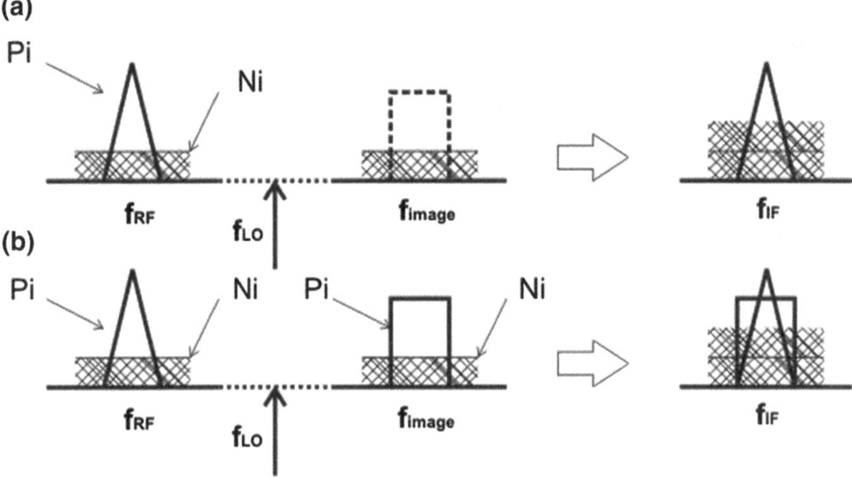

Fig. 2.33 Definition of **a** SSB versus **b** DSB NF

$$CG_v = \frac{V_{IF}}{V_{RF}} \tag{2.22}$$

$$CG_p = \frac{V_{IF}^2/R_{out}}{V_{RF}^2/R_{in}} = CG_v^2 \frac{R_{in}}{R_{out}} \tag{2.23}$$

Mixer port impedances also should be defined unless the mixer interfaces remain internal to the IC. Isolation between ports is also important. For example, LO signal leaking to the RF port can reach the receiver antenna leading to unwanted signal radiation and additional frequency sidebands through the mixing action. Port-to-port isolation can thus be defined to avoid unwanted feed-through actions in the mixer.

2.5.2 Mixer Topology

Based on the way mixing is performed, mixers can be divided into three categories: single-ended, singly-balanced and doubly-balanced mixers [3]. Single-ended mixers depend on system non-linearity to generate second-order terms resulting in mixing behavior. This can be implemented using a single MOS transistor, which is characterized by the square law I-V behavior in saturation mode. Single-balanced mixers depend on multiplication in current domain to perform the mixing action [33]. One input (usually the RF signal) is single-ended and the (the LO signal) other is used differentially.

Double-balanced mixers use both input signals differentially to provide better port-to-port isolation. Active implementation of the double-balanced mixer employ two single-balanced mixers combined together. As shown in Fig. 2.34, the RF signal is first converted to current in a transconductor. The LO signal is then used to

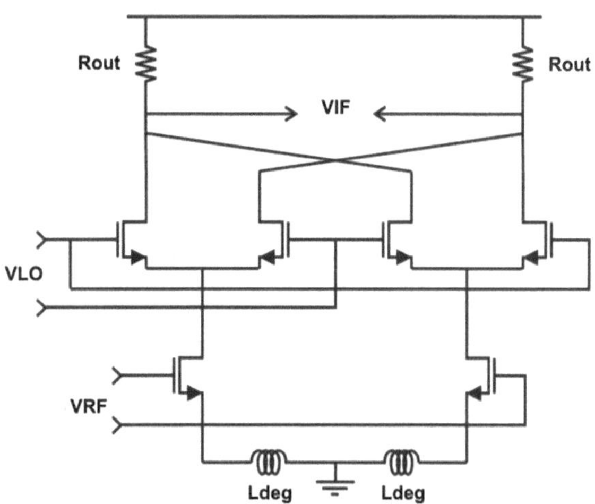

Fig. 2.34 Active implementation of double-balanced mixer

drive the switching transistors. This is equivalent to multiplying the RF current signal with a square wave that depends on the LO signal. The LO signals enter the switching transistors in anti-parallel configuration, which allows the cancellation of all related LO components at the output The fundamental frequency of the square wave, multiplied by the RF signal, will generate the required difference signal after low-pass filtering. The conversion gain for a square wave input can be calculated as following:

$$CG_v = \frac{g_m \times \frac{4}{\pi} \times R_{out}}{2} = \frac{2}{\pi} g_m R_{out} \quad (2.24)$$

where g_m is the transconductance of the RF transistor. The magnitude of fundamental component of the square wave is 4/π, and the factor of 2 is because only the difference component at the output is considered.

If the LO signal is not large enough to switch the transistors, the conversion gain will be proportional to the LO input voltage. Very high LO swings can cause the switching transistors to go into the triode regime, leading to a degraded signal path for the RF input, and so a decreasing conversion gain.

Passive implementation of double-balanced mixers provides lower noise and higher linearity with the disadvantage of having conversion loss. As shown in Fig. 2.35, a CMOS passive mixer can be enhanced with an input g_m stage and an output Op-Amp stage to provide conversion gain [3]. The input node of the Op-Amp stage is settled at virtual ground, and the LO transistors work in triode region. If the LO signal causes the transistors to switch on and off, the mixer conversion gain can be, ideally, the same as that of the active one (Eq. 2.24). The output stage is a differential Op-Amp stage with resistive feedback, which has a limited bandwidth that can work as a LPF letting only the wanted difference signal to appear at the IF output.

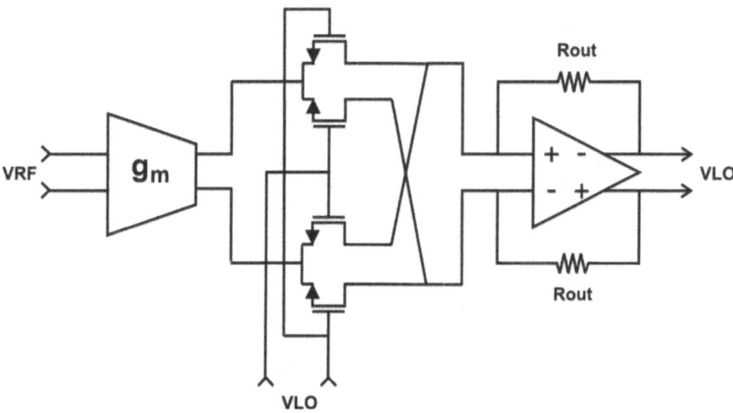

Fig. 2.35 Passive implementation of double-balanced mixer [3]

References

1. Hajimiri A, Lee TH (1999) Design issues in CMOS differential LC oscillators. IEEE J Solid-State Circ 34:717–724
2. Hegazi E, Rael J, Abidi A (2005) The designer's guide to high-purity oscillators. Springer, Berlin
3. John R, Long RF (2009) Integrated circuit design, TU-Delft. MSc course lecture notes
4. Leeson DB (1966) A simple model of feedback oscillator noise spectrum. Proc IEEE 54: 329–330
5. Craninckx J, Steyaert M (1995) Low-noise voltage-controlled oscillators using enhanced LC-tanks. IEEE Trans Circ Syst II Analog Digit Sig Process 42:794–804
6. Hajimiri A, Lee TH (1998) A general theory of phase noise in electrical oscillators. IEEE J Solid-State Circ 33:179–194
7. Huang Q (1998) On the exact design of RF oscillators. In: Proceedings of the IEEE. 1998 custom integrated circuits conference, May 1998
8. Demir A, Mehrorta A, Roychowdhury J (2000) Phase noise in oscillators: a unifying theory and numerical methods for characterization. IEEE Trans Circ and Syst-I Fundam Theory Appl 47(5):655–674
9. Rael JJ, Abidi AA (2000) Physical processes of phase noise in differential LC oscillators. In: Proceedings of the IEEE custom integrated circuits conference (CICC), pp 569–572
10. Ham D, Hajimiri A (2001) Concepts and methods in optimization of integrated LC VCOs. IEEE J Solid-State Circ 36:896–909
11. Andreani P, Fard A (2006) More on the 1/f2 phase noise performance of CMOS differential-pair LC-tank oscillators. IEEE J Solid-State Circ 41(12):2703–2712
12. Darabi H, Abidi AA (2000) Noise in RF CMOS mixers: a simple physical model. IEEE J Solid-States Circ 35(1):15–25
13. Hegazi E, Abidi AA (2003) Varactor characteristics, oscillator tuning curves, and AM–FM conversion. IEEE J Solid-State Circ 38(6):1033–1039
14. Levantino S, Samori C, Zanchi A, Lacaita AL (2002) AM-to-PM conversion in varactor-tuned oscillator. IEEE Trans Circ Syst-II Analog Digit Sig Process 49(7):509–513
15. Razavi B (1998) RF microelectronics. Prentice Hall, Englewood
16. Cheng K-W et al (2009) A gate-modulated CMOS LC quadrature VCO. In: IEEE radio frequency IC symposium, pp 267–270, June 2009
17. Mazzanti A, Andreani P (2008) Class-C harmonic CMOS VCOs, with a general result on phase noise. IEEE J Solid-State Circ 43(12):2716–2729
18. Vidojkovic V, Mangraviti G (2010) Design of 60 GHz CMOS receiver and power amplifier in 40 nm LP CMOS: LO buffer, deliverable 60 GHz-B2.2, IMEC SSET-WL, May 2010
19. Alioto M, Palumbo G (2005) Model and design of bipolar and MOS current-mode logic: CML, ECL and SCL digital circuit. Springer, Berlin
20. Suárez A (2009) Analysis and design of autonomous microwave circuits. Wiley-IEEE Press, New York
21. Rateg HR, Lee TH (1999) Superharmonic injection-locked frequency dividers. IEEE J Solid-State Circ 34(6):813–821
22. Wu H, Hajimiri A (2001) A 19 GHz, 0.5mW, 0.35um CMOS frequency divider with shunt-peaking locking-range enhancement. In: IEEE ISSCC digest of technical papers, pp 412–413
23. Tiebout M (2004) A CMOS direct injection-locked oscillator topology as high-frequency low-power frequency divider. IEEE J Solid-State Circ 39(7):1170–1174
24. Yamamoto K, Fujishima M (2004) 55 GHz CMOS frequency divider with 3.2 GHz locking range. In: Proceedings of ESSCIRC, pp 135–138, Sept 2004
25. Yamamoto K, Fujishima M (2006) 70 GHz CMOS harmonic injection-locked divider. In: IEEE ISSCC digest of technical papers, pp 600–601
26. Chen C-C, Tsao H-W, Wang H (2009) Design and analysis of CMOS frequency dividers with wide input locking ranges. IEEE Trans Microw Theory Tech 57(12):3060–3069

27. Ellinger F et al (2004) 30–40-GHz drain-pumped passive-mixer MMIC fabricated on VLSI SOI CMOS technology. IEEE Trans Microw Theory Tech 52(5):1382–1391
28. Wu C-Y, Yu C-Y (2007) Design and analysis of a millimeter-wave direct injection-locked frequency divider with large frequency locking range. IEEE Trans Microw Theory Tech 55 (8):1649–1658
29. Chien J-C, Lu L-H (2007) 40 GHz wide-locking-rangeregenerative frequency divider and low-phase-noise balanced VCO in 0.18 μm CMOS. IEEE ISSCC digest of technical papers, pp 544–545
30. Mizuno M et al (1996) A GHz MOS adaptive pipeline technique using MOS current-mode logic. IEEE J Solid-State Circ 31(6):784–791
31. Usama M, Kwasniewski TA (2004) New CML latch structure for high frequency Prescaler design. In: Canadian conference on electrical and computer engineering 2004, vol 4, pp 1915–1918, May 2004
32. Usama M, Kwasniewski TA (2006) A 40 GHz frequency divider in 90-nm CMOS technology. In: Proceeding of international north east workshop on circuits and systems (NEWCAS 2006), June 2006
33. Lee TH (2004) The design of CMOS radio-frequency integrated circuits, 2nd edn. Cambridge University Press, Cambridge
34. de Vreede L (2009) Microwave circuit design, TU-Delft. MSc course lecture notes

Chapter 3
Design and Simulation Results

Schematic simulations at 60 GHz allow the designer to become familiar with a circuit and understand its behavior and changes with different parameters. The pre-layout schematic doesn't really represent actual component values in the final design, and a considerable difference can be expected between measurements and schematic simulation results. Parasitic capacitance and inductance due to interconnects, for instance, are comparable to the designed values but cannot be accurately predicted before post-layout simulations. The design procedure followed in this work includes a few iterations. Firstly, circuit behavior is understood through schematic simulations. Initial design values can be chosen with the help of estimated interconnect parasitic capacitance and realistic values of quality factor for the passive components. A physical layout can then be drawn and post-layout results analyzed. Active and passive component values are then optimized, the layout is modified and post-layout results are again analyzed. This process is repeated few times until an optimum design is reached.

The circuit is designed in UMC 90 nm CMOS technology with thick (3.25 μm) top metal. The supply voltage is determined to be 1 V in order to have the opportunity to reuse the circuit in a 40 nm process.

In this chapter, an evaluation of the active and passive elements used in the technology is first shown. Schematic simulation results for all of the circuit blocks are then presented. This includes the QVCO, LO buffer, divider chain, LNA and mixer blocks.

3.1 f_T of the 90 Nm NMOS Transistor

Transit frequency (f_T) of a transistor determines the region of operation according to the frequency used. f_T can be calculated using the hybrid parameters. A NMOS transistor is tested at 60 GHz using the following equation:

$$f_T = 60e9 \times h_{21} \tag{3.1}$$

K. Khalaf et al., *Data Transmission at Millimeter Waves*,
Lecture Notes in Electrical Engineering 346, DOI 10.1007/978-3-662-46938-5_3

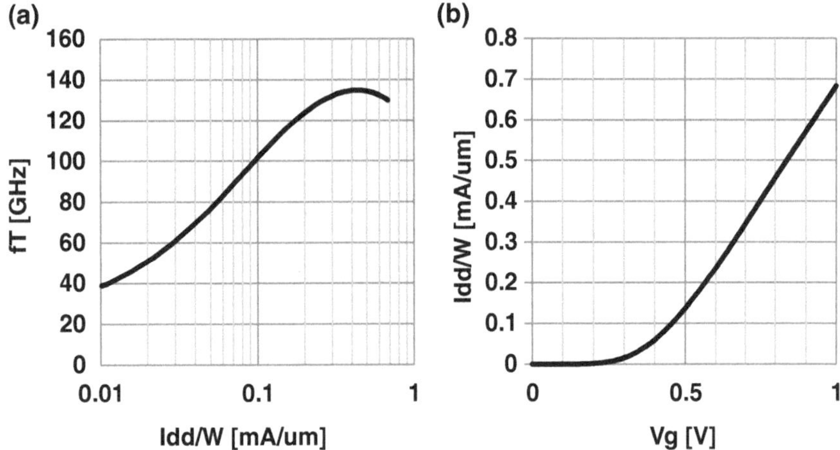

Fig. 3.1 **a** Transit frequency of the used 90 nm transistor, and **b** drain current density versus gate voltage

where h_{21} is the short circuit forward current gain. Transit frequency versus the drain current density of the NMOS transistor at minimum (80 nm) gate length is shown in Fig. 3.1. Maximum transistor f_T is 135 GHz, which is close to the operating frequency. This shows that operation at 60 GHz using this process is possible, but high transistor intrinsic gain (gm × ro) is not expected.

3.2 Passive Elements

Passive elements used in the circuit include poly resistors, diodes, metal-oxide-metal (MoM) capacitors, varactors, transmission lines, inductors and transformers. Resistors, diodes and MoM capacitors are well characterized in the available technology. Inductors and transformers are only characterized up to 20 GHz. Furthermore, not all of the values needed in the design are covered by the available components. Thus, varactors, transmission lines, inductors and transformers are designed independently, and presented in the following sub-sections. ADS-Momentum® [1] is used to simulate all the inductive elements.

3.2.1 Varactors

Inversion-mode MOS varactors are used in this design. Capacitance variation and quality factor with tuning voltage for the minimum length varactor are shown in Fig. 3.2a, b, respectively. The finger width (W) used is 1 μm and 20 fingers (i.e., M = 20) are used for each transistor. Measured quality factor is expected to be

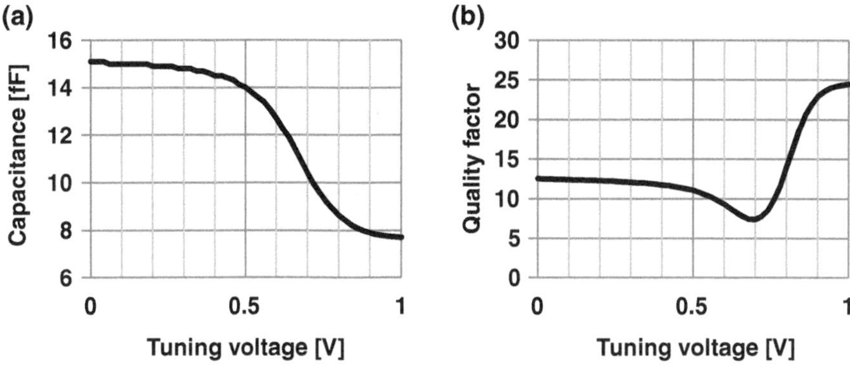

Fig. 3.2 **a** Capacitance and **b** quality factor for a varactor with W = 1 μm and M = 20

lower than the simulated value due to the lack of an accurate model at 60 GHz and the presence of interconnect parasitic resistance. Thus, margin should be added to the design or additional resistance should be added to the schematic in order to account for the reduced quality factor.

3.2.2 Transmission Lines

Coplanar waveguide over ground plane are used to implement 50 Ω transmission lines (TLs). Transmission lines with 50 Ω characteristic impedance (Zc) are needed for the chip output signals at high frequencies. Single-ended transmission lines are going to be used in the QVCO and divider subsystem of Sect. 4.2, which has output signals at 30 GHz. The TL configuration and unit-element (1 μm length) lumped RLCG model are shown in Fig. 3.3a, b, respectively. The TL is implemented using top-metal with a width (W) of 2 μm and spacing (S) of 4 μm. Zc is calculated using the following equation:

$$Z_c = Re(\sqrt{B/C}) \tag{3.2}$$

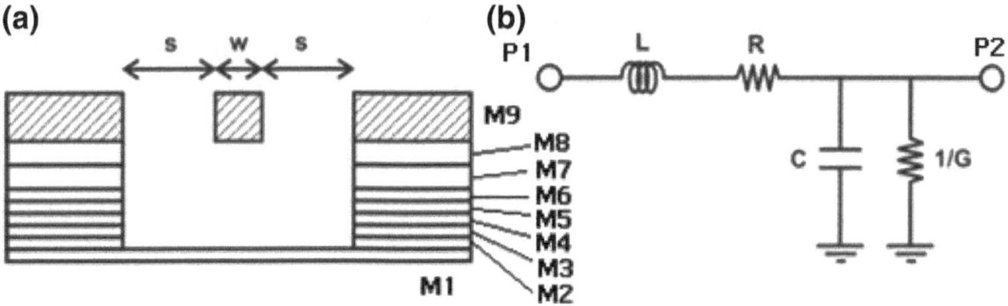

Fig. 3.3 Transmission line **a** cross-section and **b** RLCG model

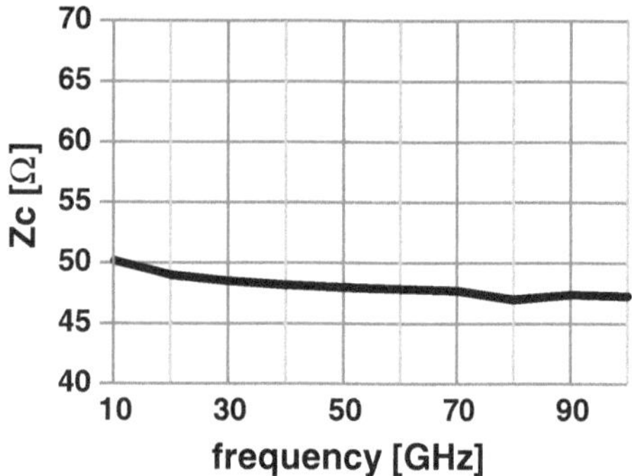

Fig. 3.4 Single-ended TL characteristic impedance versus frequency

where B and C are the transmission line ABCD-parameters. Characteristic impedance as a function of frequency is shown in Fig. 3.4. A value of 48.5 Ω was achieved for Zc at 30 GHz.

3.2.3 *Inductors*

Differential inductors are used in different blocks in our circuit. Differential inductance values from 40 to 300 pH are implemented. The smallest inductance is used in the QVCO (Sect. 3.3) and maximum size one is used in the first stage of the 60 GHz frequency divider (Sect. 3.4). A 5 × 5 μm stack of technology metal layers is used as a unit element for the chip ground plane. The square ground cell should pass the layout design-rule checks. Square inductors are implemented, shielded with ground cells during simulation to minimize the effect of other components on the inductor layout. For example, the 300 pH inductor has 2 turns with W of 3 μm, S of 2 μm and an outside dimension (OD) of 48 μm. The inductance (L) is calculated by dividing the imaginary part of the differential input impedance by the angular frequency. The quality factor (Q) is calculated by dividing the imaginary part of the differential input impedance by its real part. The simulated inductance and quality factor for the 300 pH inductor is shown in Fig. 3.5. A quality factor value of 13.3 is predicted at 60 GHz. The inductor is self oscillating at a frequency close to 100 GHz. That's why the inductance shows an increasing behavior with frequency. The self-resonance frequency is increased at lower inductor sizes due to the lower parasitic capacitance. The lumped model used in schematic simulations is shown in Fig. 3.6. Note that in this model, the total differential inductance is 2L.

3.2.4 *Transformers*

Two metal layers are used to implement the transformer. These are metal layer 9 (M9) and metal layer 8 (M8). The top metal layer (M9) is 3.25 μm thick with a

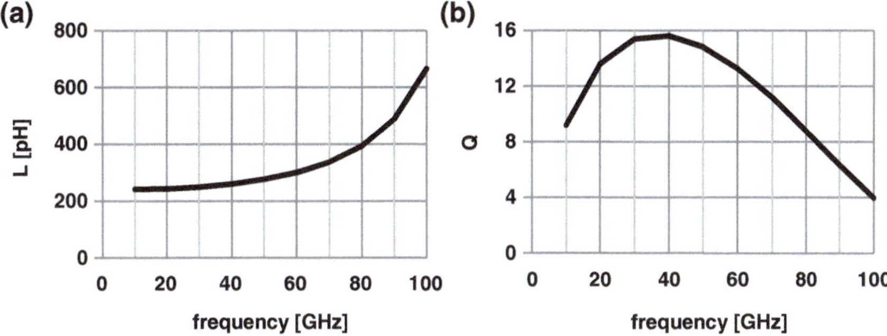

Fig. 3.5 **a** Inductance and **b** quality factor of the 300 pH differential inductor

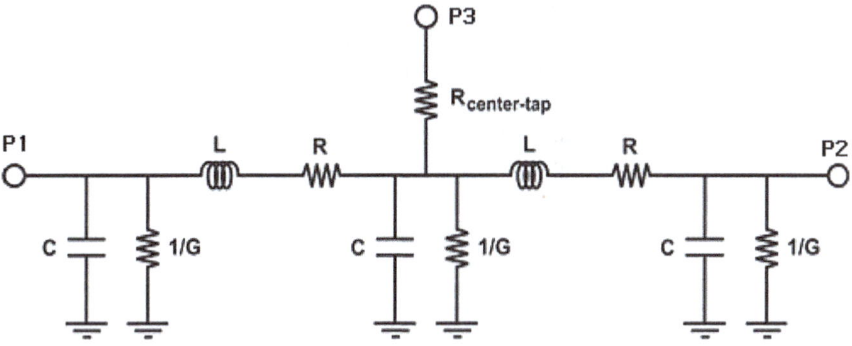

Fig. 3.6 Differential inductor lumped component model

sheet resistance of 7 mΩ/square. M8 is 0.5 μm thick with a sheet resistance of 44 m Ω/square. The inductor only uses M9. Thus, Transformers are usually implemented with lower quality factors compared to inductors. Figure 3.7 shows an 83 pH transformer. The transformer turns ratio is 1:1, so the primary and secondary inductance values are the same. The 83 pH transformer is implemented with two

Fig. 3.7 Transformer implemented with 83 pH primary (=secondary) inductance in 90 nm process

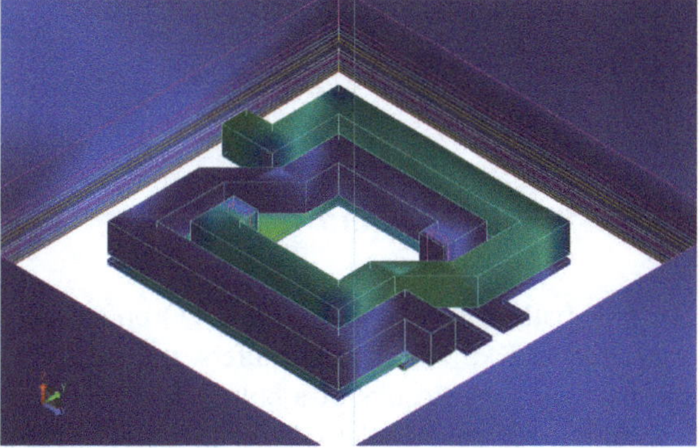

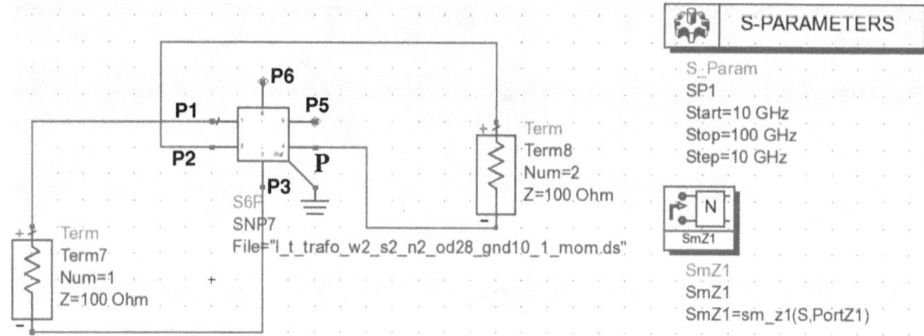

Fig. 3.8 Testbench used for the prediction of transformer parameters

turns, W of 3 μm, S of 2 μm and OD of 28 μm. This is the smallest size transformer implemented in this work, which is used in the LNA and mixer combination (see Sect. 3.5). The test-bench used to predict the transformer parameters is shown in Fig. 3.8. The mutual inductance (M) is calculated by dividing the imaginary part of Z21 (numbers refer to the differential ports in the test-bench) by the angular frequency. The coupling coefficient (k) is calculated as:

$$k = M/\sqrt{L_1 L_2} \tag{3.3}$$

where L_1 and L_2 are the imaginary parts of Z11 and Z22 divided by the angular frequency, respectively. Simulation results are shown in Fig. 3.9. A quality factor of 7.91 and coupling coefficient of 0.73 are achieved at 60 GHz. The lumped model used in schematic simulations is shown in Fig. 3.10.

3.3 QVCO and LO Buffer

The aim of this section is to design a quadrature VCO and LO buffer stage. Target specifications for the QVCO and LO buffer blocks are shown in Table 3.1. More emphasis was put on the phase noise spec by the system designer. So, a low phase noise (less than −90 dBc/Hz) QVCO is the first priority. The starting point of target phase noise spec is the −85 dBc/Hz achieved in [2].

3.3.1 Circuit Schematic

The schematics of the P-QVCO and LO buffer are shown in Figs. 3.11 and 3.12, respectively. Two similar LO buffers are used for both in-phase and quadrature oscillator outputs. Results of a bottom-series QVCO (BS-QVCO) are going to be compared with the parallel QVCO (P-QVCO). The schematic of the BS-QVCO

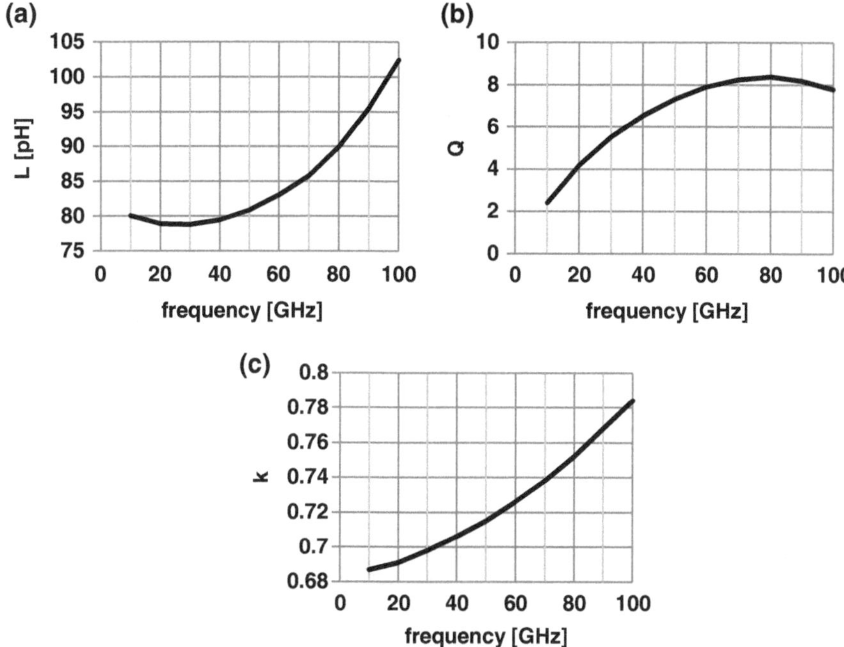

Fig. 3.9 **a** Inductance, **b** quality factor and **c** coupling coefficient of the 83 pH transformer

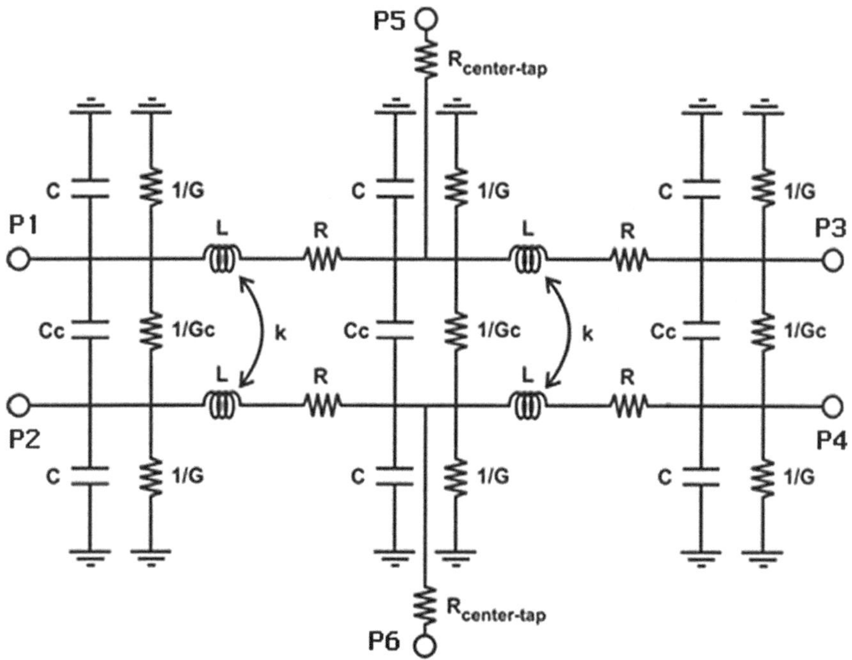

Fig. 3.10 Transformer lumped component model

Table 3.1 QVCO and LO buffer target specifications

Parameter	Value
Maximum phase noise	−90 dBc/Hz @ 1 MHz offset
Center frequency	60.5 GHz
Minimum tuning range	8 GHz
Maximum power consumption	30 mW
Minimum voltage swing	1 V-pp (rail-to-rail)

used is shown previously in Fig. 2.15b. Unless otherwise specified, the MOSFET finger width is 1 μm.

3.3.2 Circuit Operation

As shown in Fig. 3.11, a modified P-QVCO can be used to lower the phase noise. The circuit uses external gate bias for the active cross-coupled transistor pair to improve the phase noise performance [3].

3.3.2.1 External Gate Bias

In a normal cross-coupled LC VCO, the gate of one transistor is connected to the drain of the other, representing the two differential VCO outputs. This can easily take the transistor out of saturation when the peak differential output signal goes above the threshold voltage (in this case, condition of a transistor being in saturation is $Vg - Vd < Vth$). When the gate bias is independently reduced, gate-to-drain maximum voltage is reduced, allowing greater voltage swings with active transistors in saturation. This is shown graphically in Fig. 3.13.

External bias of the active cross-coupled pair is implemented through the use of gate decoupling capacitors (Cd) and biasing resistors (Rd). Cd is implemented using metal-oxide-metal (MOM) capacitors. Rd is implemented using P+ poly resistors. This can partially load the VCO output through parasitics of active transistors. However, the improved phase noise performance due to the independent gate bias encourages using it.

3.3.2.2 Modal Determinism

In a quadrature-VCO, two modes of oscillation can occur when the oscillator is switched on [4]. This depends on the quadrature signal, either leading or lacking the in-phase signal by 90°. A detailed mathematical analysis was performed on the QVCO to derive its two modes [5]. The analysis predicts a fast, high frequency,

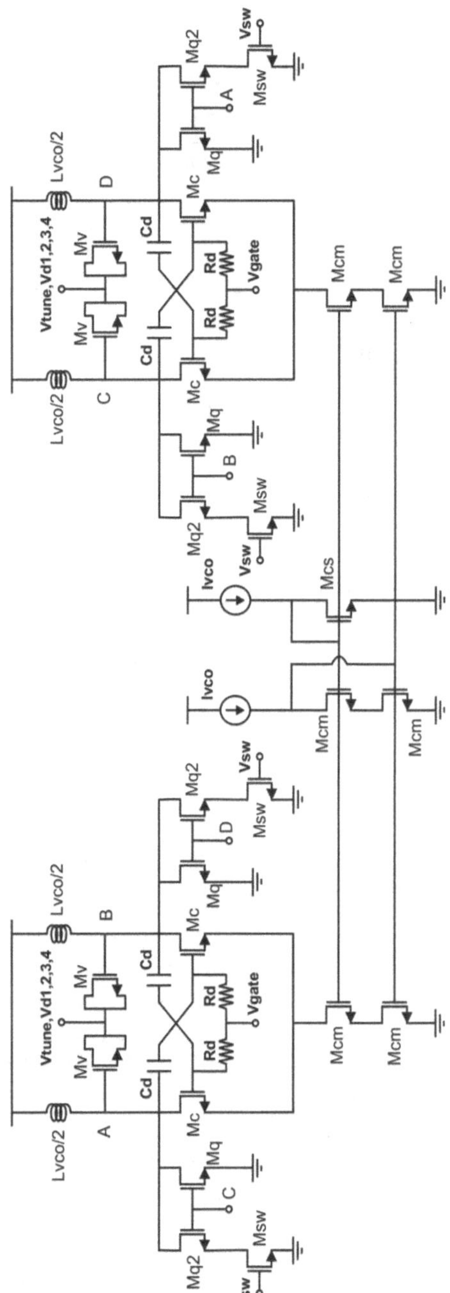

Fig. 3.11 Circuit schematic of the used P-QVCO

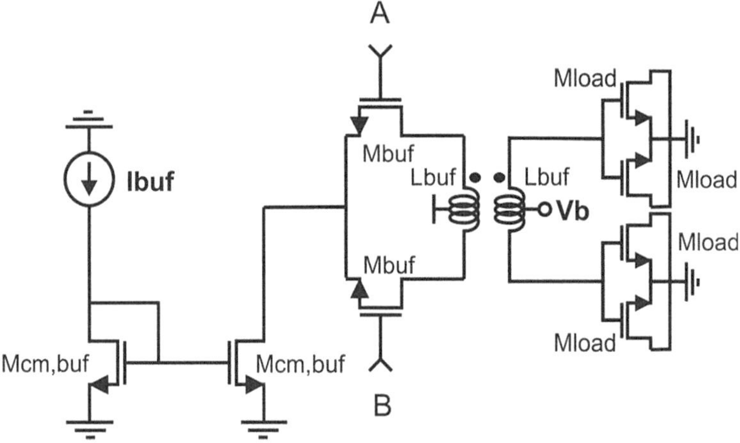

Fig. 3.12 Circuit schematic of the transformer-coupled LO buffer with output load

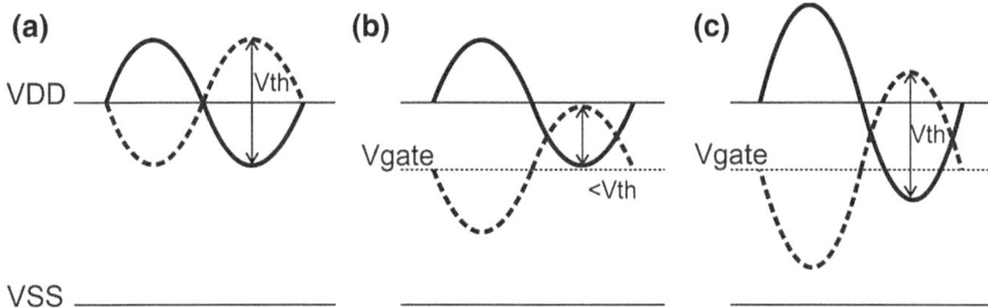

Fig. 3.13 QVCO **a** without **b** with external gate bias, and **c** higher swing possibility without taking the cross-coupled transistors out of saturation

mode and a slow mode. Owing to the asymmetry in the tank impedance due to the difference between the inductor and varactor quality factors, one frequency mode usually dominates the other. A figure illustrating the two modes' impedance points on the tank impedance curve is presented in Fig. 3.14. The QVCO tends to oscillate at the higher tank impedance mode. The difference between the two tank impedance modes should be large enough to ensure operation at one of them. A higher coupling coefficient operates the QVCO at tank impedance points with a larger difference in value. This explains the need for higher coupling coefficient to ensure modal determinism. A higher coupling coefficient also increases the effective G_m and improves the oscillation margin which is required for the oscillator to startup. However, phase noise is degraded by a higher coupling coefficient. Thus, variable coupling can be implemented to achieve all requirements. When switching on the oscillator, a high coupling coefficient is used for the startup. Then a low value is selected for lower phase noise.

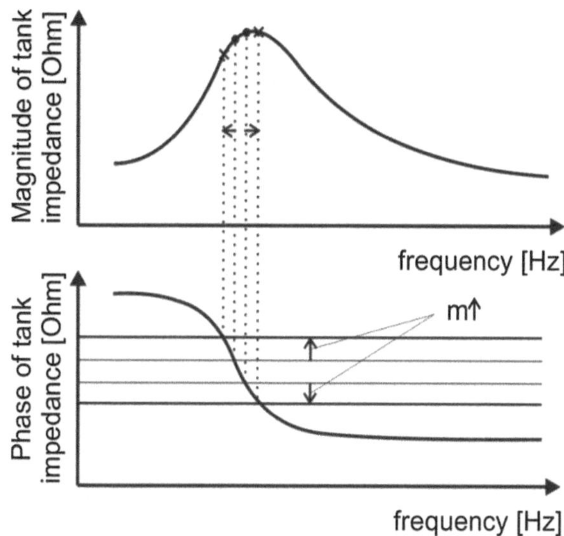

Fig. 3.14 Impedance points of the QVCO two oscillation modes [5]

3.3.2.3 Variable Coupling

Variable coupling is implemented using three transistors. One fixed small coupling transistor (Mq) and a larger switching coupling transistor (Mq2). When the switch is ON, Mq2 is in parallel with Mq1, giving a high value of coupling coefficient. Two transistors are used to implement variable coupling instead of one in order to ensure quadrature locking at all values of the switching voltage.

3.3.2.4 Digital Varactor

There is a tradeoff between tuning range and phase noise using a single varactor. A digitally controlled varactor using a bank of binary weighted sizes can be used. As mentioned earlier in Sect. 2.1.3, noise in active elements can easily be transformed into phase noise by the tank varactor due to its sensitivity to amplitude variations. The oscillator's sensitivity to the varactor can be reduced by using a smaller varactor size ($Kv = \Delta C/\Delta V$). However, the analog varactor should be larger than the minimum digital varactor size. This is to overcome process variations and ensure overlapping frequency ranges. This improves AM-to-FM conversion, and thus, reduces translated phase noise components in the circuit [6, 7].

3.3.2.5 Current Mirror

A cascode current mirror is used to implement a high output resistance. This is useful to reduce the noise generated in one of the cross-coupled pair transistors while the other is OFF, as mentioned in Sect. 2.1.3. Lengthening the current mirror transistors reduces flicker noise. Further increase of the tail transistor length, with its width set to the maximum value (limited by layout), to increase the output resistance causes less current copy ratio. This is because of the reduced aspect ratio (W/L), which leads to an increase in the common gate-source voltage of the current mirror transistors and the drain-source voltage of the diode-connected one. The drain-source voltage of the tail transistor, on the other hand, is controlled by the supply. Due to the limited headroom available from the 1 V supply, tail transistors are working on the edge of saturation.

3.3.3 Design Guidelines

For oscillator design at 60 GHz, the following equation should be used:

$$60 \times 10^9 = \frac{1}{2\pi\sqrt{LC}} \tag{3.4}$$

where L and C are the total inductance and capacitance values, respectively, seen at the output of the oscillator. The total output capacitance is due to several components including the cross-coupled transistor pair, parallel coupling transistors, varactor, buffer load and interconnect parasitics.

3.3.3.1 Cross-Coupled Pair

More transconductance is required from the cross-coupled transistor pair (gm,c) to start-up the oscillation with sufficient margin. However, its width (Mc) can't be increased so much to keep a room for other capacitance components to tune the resonator. Thus, a good balance between different capacitive components is required for the oscillator. Mc was, thus, chosen to be 40 μm.

3.3.3.2 Digital Varactor

Four digitally-controlled inversion-mode MOS capacitors are chosen to provide a discrete frequency step for the VCO. Sizes of the digital varactors are binary-weighted to cover the overall frequency range. The finger width of a transistor is a trade-off between input resistance and capacitance values. The finger width used in the varactor is 0.5 μm, which is the minimum finger width available from the technology. This improves varactor quality factor at the expense of lower tuning range at the same center frequency. The larger the size of the varactor is, the larger the tuning range. However, this loads the VCO output and causes a drop in the center frequency at the same inductance value. Varactor sizes of 24, 48, 96 and

192 μm, with an analog varactor of 36 μm, are used to provide a tuning range of 8 GHz centered at 60 GHz (i.e., 13.3 %). This is limited by the minimum possible inductance value.

3.3.3.3 Differential Inductor

The smallest inductor is 2 μm wide, 18 × 18 μm half-turn (i.e., U-shaped) to connect the two drains of the cross-coupled transistor pair. The minimum inductance was found to be around 40 pH with a quality factor of 14.2. An initial value of 45 pH with a quality factor of 15 was used in the schematic. A small inductor is required at 60 GHz to compensate the large capacitance value at the VCO due to the previously mentioned contributors. With the chosen inductor, transistor size and varactors, a tuning range of 8 GHz with 500 MHz step size is simulated.

3.3.3.4 Quadrature Coupling

Coupling transistor Mq should be small to represent a small coupling coefficient for improved phase noise. Minimum coupling coefficient of 0.1 is chosen. With gm,c of 30 mS, this requires that gm,q is 3 mS, which gives a value of 4 μm for Mq. A very small value for the transistor width can cause more mismatch ($\sigma \alpha 1/\sqrt{\mathrm{WL}}$) [8]. Also contribution of transistor flicker noise is high at lower transistor widths. A value of 6 μm is chosen for lower flicker noise contribution of Mq and better mismatch. This gives a minimum coupling coefficient of 0.15. Mq2 in combination with Msw give the second part of the coupling coefficient. The switching transistor works in triode, and can be replaced by a resistor. The effective coupling transconductance when the switch is ON can be written as:

$$gm{,}q{,}eff = gm{,}q + \frac{gm{,}q2}{1 + gm{,}q2 \times Rsw} \tag{3.5}$$

where gm,q,eff is the total effective coupling transconductance, gm,q is the fixed coupling transconductance, gm,q2 is the variable coupling transconductance and Rsw is the switch transistor equivalent resistance. Mq2 should be large to have a better margin of oscillation startup and modal determinism. A large Mq2 also adds capacitance to the QVCO output, which shifts the oscillation frequency to lower values. As Mq2 is only used to ensure appropriate startup, a maximum coupling factor of 0.7 is chosen. This leads to a value of 21 mS for gm,q,eff (this value is based on the simulator). With a large enough switching transistor to ensure low Rsw, Mq2 can be determined according to the resulting gm,q2 value. For Msw of 48 μm, Rsw is 9.7 Ω and gm,q2 is 21 mS, leading to a value of 18 μm for Mq2.

3.3.3.5 Current Mirror

Wide channel transistors are required in the current mirror in order to have the best possible current copy ratio and to decrease flicker noise. The limited headroom available from the 1 V supply and the reduced gate voltage for better phase noise performance caused a small value (around 140 mV) of Vds to be available for tail transistors. This caused around 7 mA only to be copied from a 10 mA current source (30 % current loss in the current mirror). A value of 208 μm total width was chosen for the current mirror transistors (Mcm). The maximum number of fingers of a transistor in the used technology is 32. With a finger width of 1 μm, 8 parallel transistors of 26 μm each were used to implement one 208 μm current mirror transistor. Even number of fingers was suggested in the technology for the best matched transistor layout.

3.3.3.6 Current Source

Two current sources, 10 mA each, feed the QVCO through the cascode current mirror. As explained in Sect. 2.1.2, the VCO operation is divided into current-limited and voltage-limited regimes according to the bias current. As shown in Fig. 3.15, the output amplitude is proportional to the bias current in the current-limited regime. This leads the phase noise to decrease with increasing the bias current as expected from the phase noise equation (Eq. 2.9). In the voltage-limited regime, the tail transistor goes into the triode region and the output amplitude is almost constant, limited by the power supply. Operation in the voltage-limited regime is usually not desirable, because the bias current is wasted without an effective increase in the output amplitude. Moreover, the higher current leads to a higher transconductance (higher noise current), which degrades the phase noise. Thus, the VCO should operate on the edge of the voltage-limited regime in order to have the best phase noise without wasting the bias current.

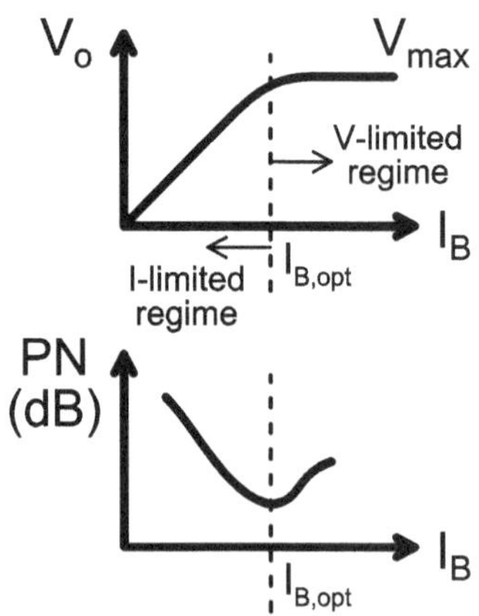

Fig. 3.15 Current and voltage-limited regimes with the bias current as a variable [9]

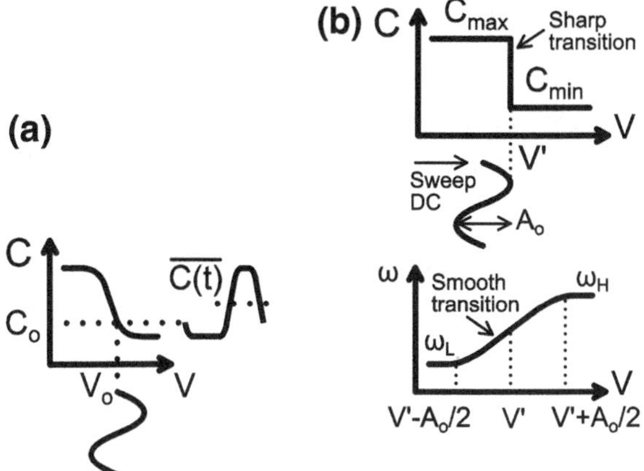

Fig. 3.16 **a** Average capacitance value and **b** effective capacitance curve for a varactor in a VCO [10]

3.3.3.7 Output Voltage Swing

One drawback of high output voltage swing in the VCO is the reduced possible tuning range [7, 10]. The capacitance value at a specific varactor bias is averaged due to the time varying oscillation amplitude, as shown in Fig. 3.16a. As shown in Fig. 3.16b, effective capacitance causes the actual tuning curve to be smoother than the DC varactor characteristics. Higher amplitude causes less tuning sensitivity, and thus, lower available tuning range. That's because available tuning voltage is limited by the supply. This effect can be shown in Fig. 3.17. Thus, output amplitude should be kept in the current-limited regime, as a compromise between phase noise and tuning range. In our circuit, maximum output swing in the QVCO (before the LO buffer) was kept around 250 mV.

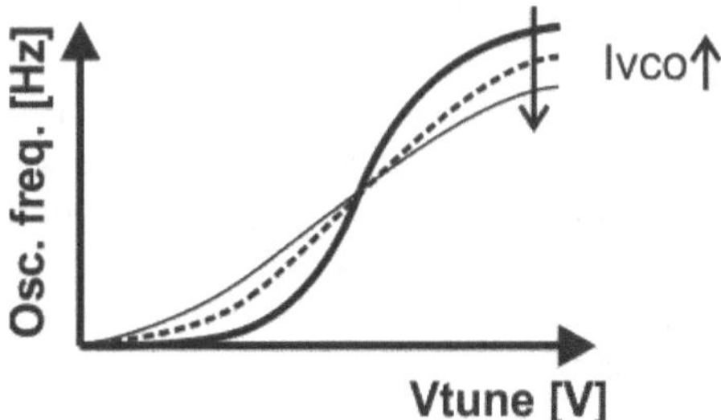

Fig. 3.17 Less tuning frequency with higher current due to different effective capacitance values at different amplitudes

3.3.3.8 External Gate Bias

The cross-coupled transistor pair gate voltage (Vgate) should be lower than the supply voltage, to keep transistors in saturation (for peak differential amplitudes higher than Vth) and improve phase noise. Vgate also controls the common source node of the cross-coupled pair, which is the headroom for the tail transistor. Low headroom can get the tail transistor out of saturation. The tail transistor will fail to work as a high impedance current source. This leads the phase noise to, again, start increasing. So, Vgate was chosen to be 0.6 V as a trade-off between phase noise and tail transistor headroom.

3.3.3.9 LO Buffer Transistor Size

The LO buffer transistor size (M_{buf}) should be kept as small as possible not to load the VCO output. However, large buffer size is required for large transconductance (gm,buf) to overcome the buffer load resistance. The LO buffer load resistance also depends on the following stage. In our circuit, we simulate using large buffer load transistors (two parallel transistors with 20 μm total width each). This is to simulate the large-size input transistors of the first-stage divider. The LO buffer gain is due to the multiplication of G_m and R_{out}. R_{out} includes the buffer transistor output resistance, which is inversely proportional to the width. So, as the buffer size is increased, G_m is increased and R_{out} is decreased. The buffer gain will, thus, increase to the point at which the reduction in R_{out} is more dominant. So, there is an optimum width for the buffer transistors to get maximum gain at a specific load. In our circuit, we chose M_{buf} to be 14 μm. This value is still below the optimum value for maximum gain. However, a compromise between buffer gain (controlling output voltage swing) and QVCO loading (controlling center frequency and tuning range) was considered.

3.3.3.10 LO Buffer Configuration

The LO buffer can either be an inductively-tuned or a transformer-coupled CS differential amplifier, shown previously in Sect. 2.2. The selection between both configurations depends on the buffer load. Capacitive part of the load can be compensated by the inductor or the transformer. Resistive part of the load, however, always contributes to the reduction of the buffer gain, and can only be overcome by the buffer transconductance. In the transformer-coupled choice, the load resistance is transferred to the buffer transistor drain terminal with a higher value. This improves the resistive contribution of the load transistor to the total buffer output resistance (see Eqs. 2.11 and 2.16). However, a transformer is usually implemented with a low quality factor, as compared to the differential inductor. In this technology, differential inductors with Q of around 15, and transformers with Q of around 8 and coupling coefficient of 0.8 were implemented. The transformer implemented for this circuit is a 160pH one with 2 turns, W of 2 μm, S of 2 μm and OD of 32 μm. Lower quality factor is equivalent to a lower parallel resistance, which degrades the resistive contribution of the inductive component to R_{out}. So, depending on the effective buffer output resistance at a specific load, a choice can be made between both configurations. In our circuit, a large 40 μm width transistor was chosen to load the LO buffer. That would degrade R_{out} if connected directly to the buffer output. Thus, a transformer-coupled CS differential amplifier was chosen to load the QVCO.

3.3.4 Design Values

All transistor sizes, passive values and controlling parameters are listed in Table 3.2. Finger width of all transistors is 1 μm except for the digital varactor which has a 0.5 μm width. M defines the number of fingers and L is the channel length.

3.3.5 Simulation Results of P-QVCO

In the following section, the schematic simulation results of the designed P-QVCO and transformer-coupled LO buffer combination are plotted. An ideal inductor with an estimated Q of 15 was used in the QVCO. An additional capacitance of 10 fF was added at each node of the circuit to simulate the parasitic capacitance that is expected to be added after layout.

3.3.5.1 Amplitude and Tuning Range

Figure 3.18 shows the P-QVCO single-ended peak output amplitude (A) over the whole tuning range. A plot of the LO buffer output voltage swing (A_{buf}) is also plotted on the same graph. A peak of around 0.42 V at the buffer output is achieved at a 40 μm transistor load. This corresponds to 0.84 V peak-to-peak, which is acceptable compared to the rail-to-rail (1 V) requirement. Note that output amplitudes increase with smaller transistor sizes for the load. So, a comparison with a smaller-size load will be shown in Sect. 3.3.5.6.

The oscillation frequency is plotted on the x-axis. It shows that a tuning range of 8 GHz centered at 60.8 GHz is achieved (from 56.8 to 64.8 GHz). The full 8 GHz are divided into 16 sub-ranges, 500 MHz each, by the four digital varactors.

3.3.5.2 Phase Noise

As shown in Fig. 3.19, the phase noise at 1 MHz offset (fd) changes from −98.9 to −94.8 dBc/Hz over the tuning range. Phase noise performance shows some degradation at higher oscillation frequencies. Apart from the reduced amplitude, the degradation in phase noise is attributed to the lower equivalent tank capacitance

Table 3.2 Design values for the QVCO and LO buffer

Transistor sizes						
Parameter			P-QVCO	Reduced load	BS-QVCO	Reduced load
Cross-coupled pair (Mc)		M	40			
		L	80 nm			
Fixed coupling (Mq)		M	6	6	30	30
		L	180 nm			
Variable coupling (Mq2)		M	18		×	×
		L	80 nm		×	×
Switch transistor (Msw)		M	48		×	×
		L	100 nm		×	×
Digital varactor (W = 0.5 μm)	Md1	M	96			
		L	80 nm			
	Md2	M	48			
		L	80 nm			
	Md3	M	24			
		L	80 nm			
	Md4	M	12			
		L	80 nm			
Analog varactor (Mv) (W = 0.5 μm)		M	18			
		L	80 nm			
QVCO Current mirror (Mcm)		M	208			
		L	200 nm			
QVCO Current source (Mcs)		M	72			
		L	200 nm			
Buffer transistor (M_{buf})		M	14			
		L	80 nm			
Buffer current mirror (Mcm, buf)		M	208			
		L	300 nm			
Load transistor (Mload)		M	20	12	20	12
		L	80 nm			
Passive elements' values						
QVCO differential inductor		Lvco	45 pH			
Gate decoupling capacitor		Cd	540 fF			
Gate biasing resistor		Rd	3 kΩ			
Buffer transformer (primary = sec.) (pH)		L_{buf}	160	230	160	230
Controlling parameters						
External gate voltage		Vgate	0.6 V		×	×
QVCO bias current (mA)		Ivco	10	10	5	5
Buffer bias current		I_{buf}	5 mA			

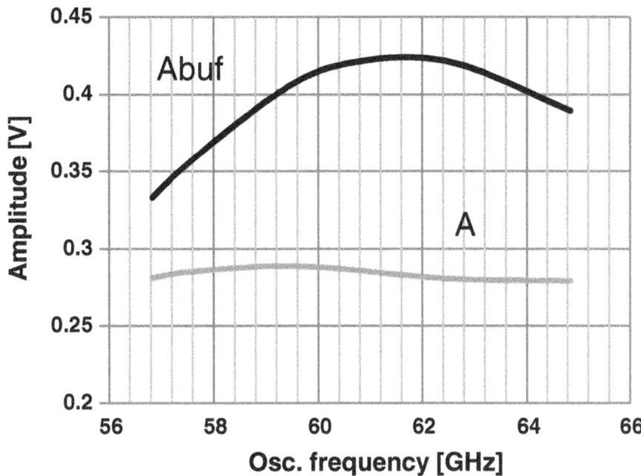

Fig. 3.18 P-QVCO and LO buffer amplitudes versus oscillation frequency

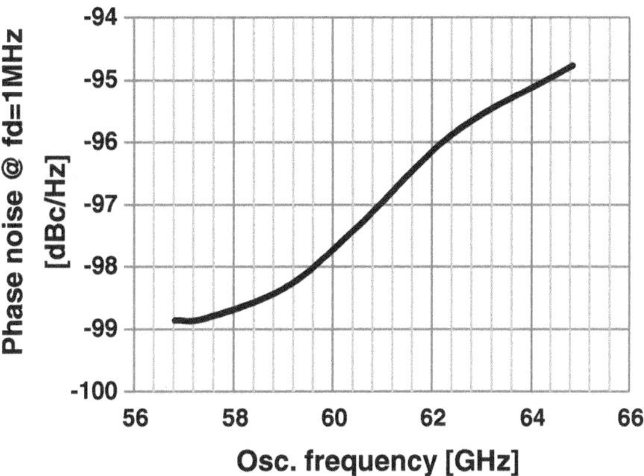

Fig. 3.19 Phase noise at 1 MHz offset versus oscillation frequency

(higher oscillation frequency) as expected from Eq. 2.9. Moreover, the buffer input resistance, which loads the oscillator tank, is reduced due to the feedback caused by the gate-drain capacitance (Cgd,buf) of the buffer transistor (M_{buf}). One way to improve the phase noise is to use neutralization capacitors [11] to cancel the effect of Cgd. This improves the oscillator output amplitude and phase noise. This will be shown in Sect. 4.2.

3.3.5.3 Variations with Tail Current

The overall circuit operation was optimized at a 10 mA tail current. Figure 3.20a shows the amplitude increase with the QVCO bias current (Ivco). The voltage-limited regime can be recognized at around 20 mA. Operation at 10 mA is chosen to meet both power consumption and tuning range specifications. As mentioned in Sect. 3.3.3.7, tuning range is reduced with higher voltage swings (higher bias current at the same circuit design values). This can be shown in Fig. 3.20b.

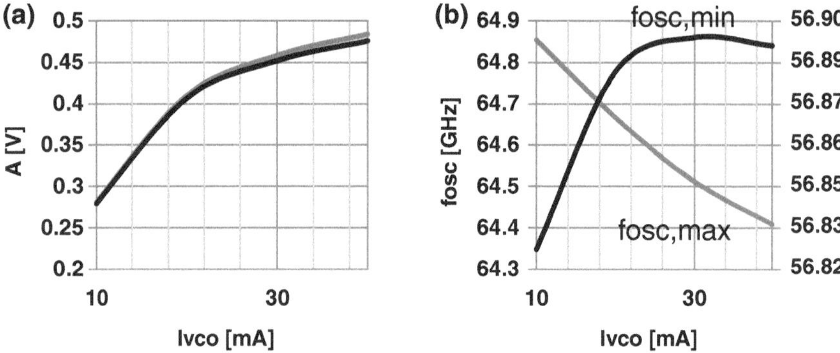

Fig. 3.20 **a** A at max. and min. oscillation freq. **b** Max. and min. oscillation frequency variations with QVCO bias current

Fig. 3.21 Phase noise variations at max. and min. oscillation frequencies with QVCO bias current

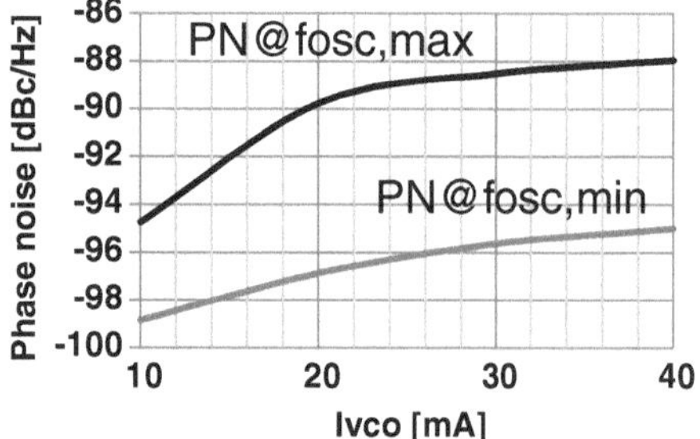

Figure 3.21 shows phase noise degradation with higher bias current. The phase noise is best at the beginning of the voltage-limited regime in Fig. 3.15. This is the AM-to-FM phase noise component. In our circuit, the overall phase noise is optimized at 10 mA. Phase noise keeps degrading in the voltage-limited regime. In that region, current is consumed without an effective increase in the amplitude. Higher current causes an increase in transistor gm. This leads to higher contribution of a transistor noise to the overall phase noise.

3.3.5.4 Variations with Gate Voltage

Circuit performance with different gate voltages is shown in Fig. 3.22. More gate bias increases the common-source node voltage of the cross-coupled pair. When this leads to an increase in the tail current (due to higher Vds), output amplitude is increased. Total phase noise is minimized at 0.6 V. Higher gate voltage takes the cross-coupled pair out of saturation. Lower gate voltage takes the tail transistor out of saturation. Both leads to an increase in phase noise as mentioned in Sect. 3.3.2.

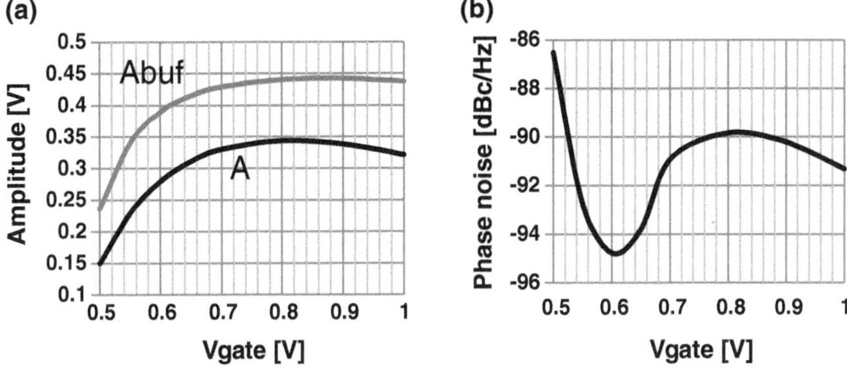

Fig. 3.22 **a** Amplitude and **b** phase noise variations with external gate bias at maximum oscillation frequency

3.3.5.5 Variations with Supply Voltage

Supply voltages up to 1.2 V are available for this process. Figure 3.23 shows how the circuit behaves at different supply voltages (all other parameters are constant). Lower supply voltages lead to lower headroom for the tail transistor. This leads to lower tail current and, thus, lower amplitudes. Phase noise is expected to increase with reduce voltage swings as clear from the plot. At 1.2 V supply, amplitude is also reduced. This is because the gate voltage is kept constant while the common-source node is increased. This leads to a reduction in the downward gate swing before switching the cross-coupled pair transistor off.

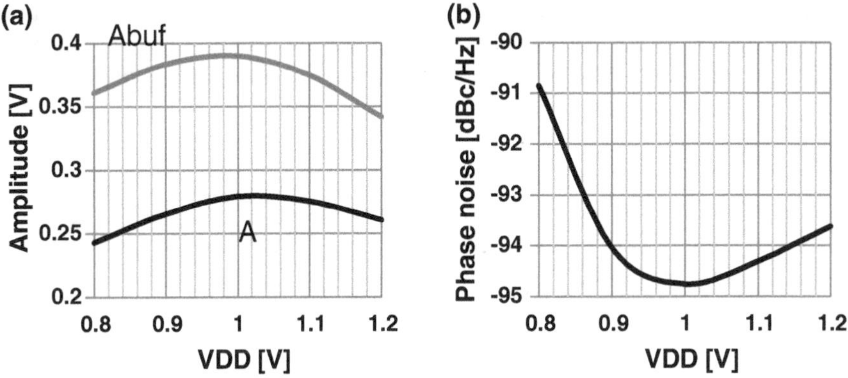

Fig. 3.23 **a** Amplitude and **b** phase noise variations with supply voltage at maximum oscillation frequency

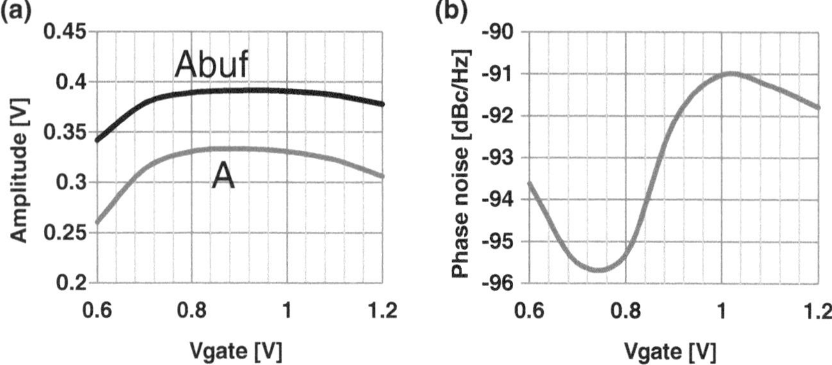

Fig. 3.24 **a** Amplitude and **b** phase noise variations with external gate bias at 1.2 V supply and maximum oscillation frequency

Low gate bias prevented the QVCO performance to be improved at 1.2 V supply voltage. So, a graph of amplitude and phase noise versus gate voltage at 1.2 V supply is plotted in Fig. 3.24. This shows an increase in output amplitude at higher gate voltage, and an improved phase noise of −95.8 dBc/Hz at Vgate = 0.75 V.

3.3.5.6 Performance at a Reduced Load Size

Up till now, the LO buffer is assumed to have a 40 μm load transistor. This is to simulate a large locking range divide-by-two stage, as will be shown in the divider section. Smaller load transistor size helps getting a better output voltage swing. The LO buffer is usually loaded by two mixer blocks in a receiver front-end, each with around 12 μm input transistor size, as will be shown in the mixer section. Figure 3.25 shows the increase in the buffer output swing at a load transistor of 24 μm instead of 40 μm. Maximum output swing is around 0.52 V peak.

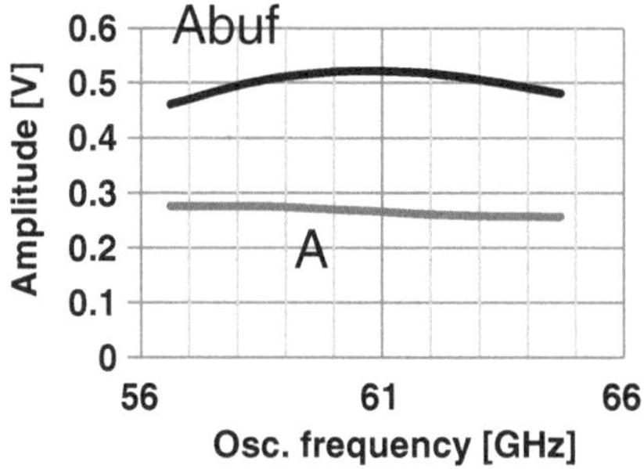

Fig. 3.25 P-QVCO and LO buffer amplitudes versus oscillation frequency at a reduced load transistor width of 24 μm

3.3.6 Simulation Results of BS-QVCO

In the following section, the schematic simulation results of the BS-QVCO and transformer-coupled LO buffer combination are plotted. The BS-QVCO is designed using the same parameter values of the P-QVCO except for the coupling transistor size and the bias current. They are optimized for the best performance of the oscillator.

3.3.6.1 Performance at a Load Transistor of 40 μm

Figure 3.26 shows the BS-QVCO performance. Peak amplitude of around 0.39 V and a tuning range of 8 GHz were achieved. At the highest oscillation frequency, phase noise is −95.1 dBc/Hz. This is due to the buffer tail transistor, as explained before.

3.3.6.2 Performance at a Load Transistor of 24 μm

At a reduced load transistor size, the BS-QVCO can reach peak amplitude of 0.49 V. This is expected because of the 100 mV improvement in the P-QVCO output amplitude when a 24 μm load transistor is used instead of a 40 μm one (Figs. 3.18, 3.25 and 3.27).

3.3.7 Performance Summary

A comparison between the required and achieved specs in both the P-QVCO and BS-QVCO is shown in Table 3.3. All the requirements are met from simulations of the circuits at the schematic level. The output swing requirement is met at a reduced

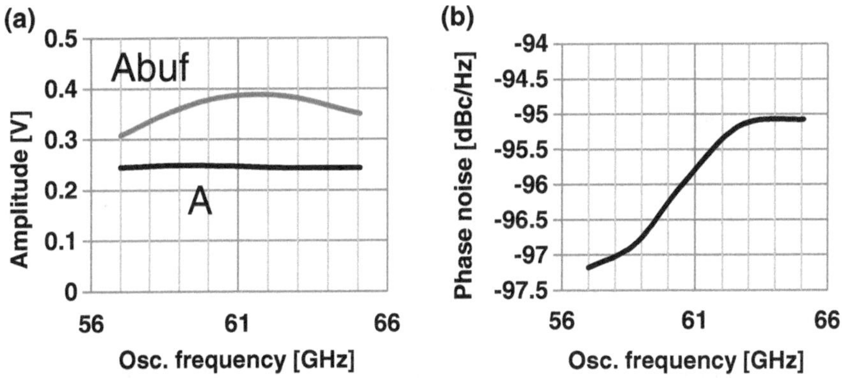

Fig. 3.26 BS-QVCO and LO buffer **a** amplitudes and **b** phase noise over the whole tuning range

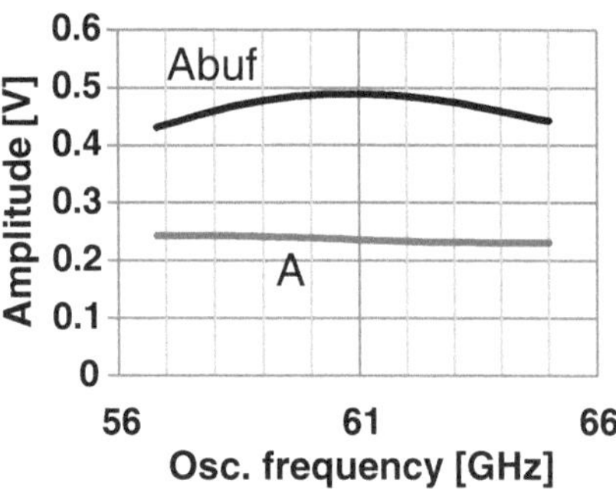

Fig. 3.27 BS-QVCO and LO buffer amplitudes versus oscillation frequency at a reduced load transistor width of 24 μm

Table 3.3 QVCO and LO buffer target specs and achieved results

Parameter	Required	Achived (P-QVCO)	Achieved (BS-QVCO)
Max. phase noise	−90 dBc/Hz	−98.9 to −94.8	−97.2 to −95.1
Center frequency	60.5 GHz	60.8 GHz	60.5 GHz
Min. tuning range	8 GHz	8 GHz	8 GHz
Max. power consumption	30 mW	26.6 mW	10 mW
Min. voltage swing	1 V-pp	0.84 V-pp (Mload = 40 μm)	0.78 V-pp (Mload = 40 μm)
		1.4 V-pp (Mload = 24 μm)	0.98 V-pp (Mload = 24 μm)

load size of 24 μm. The BS-QVCO shows very close results to the externally gate-biased P-QVCO. The advantage in the series approach of the BS-QVCO is the current reuse. Power consumption is 10mW in the series-QVCO compared to the 26.6mW in the parallel one. The P-QVCO was selected due to the variable coupling, which ensures oscillation startup. Thus, the P-QVCO configuration will be used in the following top-level circuits.

3.4 Divider Chain

The aim of this section is to design a divider chain with a divide ratio of 16, with maximum input locking range, and minimum power consumption. This ratio will divide the 60 GHz input to low-GHz frequencies (3.75 GHz), which is easily measured by an oscilloscope. Also large divider ratios can be reused afterwards in a phase-locked loop (PLL) design, which requires a divider block in its feedback path.

A combination of analog and digital dividers was used to provide a reasonable area and power consumption with good characteristics. A chain of four divider blocks, each with a divide-by-two, is chosen to divide the 60 GHz signal to low-GHz frequencies. Multiple divide-by-two blocks are easier to implement and less complex compared to other higher order dividers. Static dividers don't use inductors, and are thus smaller in size. However, the maximum frequency of operation in static dividers is proportional to power consumption. So, for high frequencies (down to 30 GHz input), analog dividers are used. The first two blocks are then analog dividers, and the last two are digital ones. Only one inductor per analog divider is used to have a compact design.

Special care should be taken to provide enough margin in the locking range of each divider to overcome process variations and keep the input and output frequency ranges matched within all the divider blocks.

3.4.1 Circuit Schematic

The first two stages of the divider chain are injection-locked frequency dividers (ILFDs), each with two injecting transistors connected across the tank. The dual-mixing direct ILFD is explained in Sect. 2.3.1, and its schematic is shown in Fig. 2.26. The second two stages use a CML static divider, with its schematic shown in Fig. 2.28. The complete divider block diagram is shown in Fig. 3.28, and the 50 Ω output buffers are shown in Fig. 3.29.

3.4.2 ILFD Locking Range

Equation 2.17 is an expression for the locking range of direct ILFDs, derived as a result of the analytical model developed in [12]. The equation shows that the locking range doesn't depend on the tank quality factor (Q_{tk}). Q_{tk} can only affect the locking range indirectly through the output amplitude. For example, lower Q_{tk}

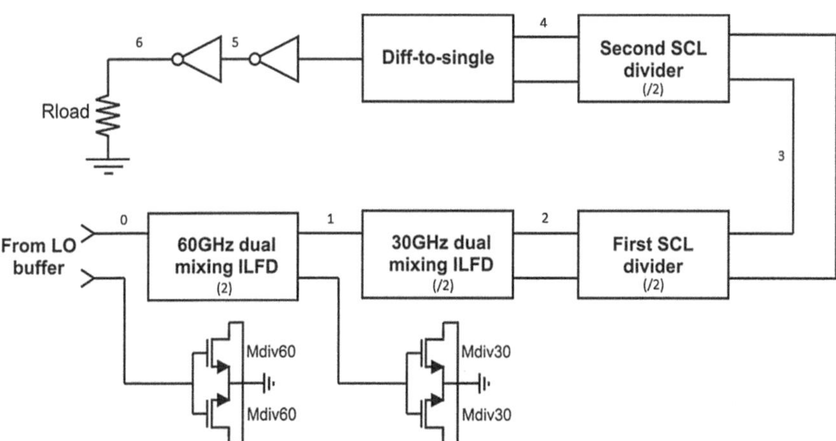

Fig. 3.28 Divider chain block diagram

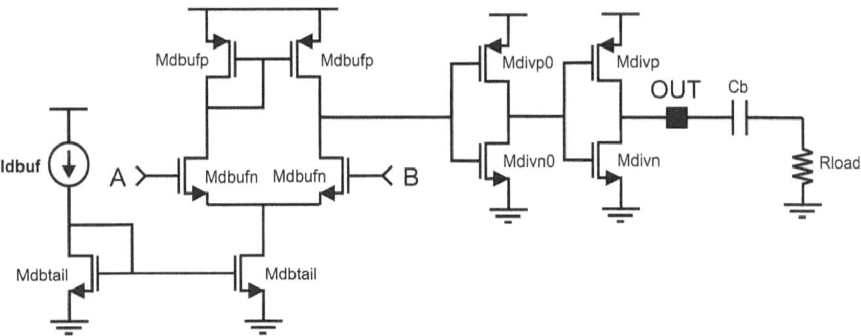

Fig. 3.29 Differential to single-ended stage with output buffers

results in a lower output amplitude, which leads to a higher equivalent injecting transistor output conductance ($g_{q,max}$) and higher locking range. However, a very high (e.g., 1000) Q_{tk} can only pass a narrow frequency range. A divider using this load cannot have the same locking range as another one using a low (e.g., 10) Q_{tk} with wide bandwidth. In this section, we're going to show that the locking range is independent of Q_{tk} only to a maximum value of the tank quality factor ($Q_{tk,opt}$). The locking range after this $Q_{tk,opt}$ is going to decrease with increasing Q_{tk}.

To understand the effect of the Q_{tk} on the locking range at the same input overdrive voltage and output voltage swing (same $g_{q,max}$), simulations were performed on the first stage divider (Fig. 2.26). The divider is loaded only by the input stage of the following divider and 10 fF additional capacitance to model layout parasitics. All the values used are the same as the 60 GHz divider values in Table 3.4 except for Mdiv60 which is 26 μm. This is just to account for the lower load capacitance due to the usage of only the input stage of the 30 GHz divider.

With a fixed inductance value and transistor sizes, the inductor quality factor (Q) is varied between 5 and 5000 (Q represents Q_{tk} with a fixed capacitance quality factor). The tail current is adjusted at each run to have the same output voltage swing at 60 GHz. Figure 3.30 shows how the locking range changes with different Q values. The tank quality factor is the parallel combination of the inductor and capacitor quality factors. So, at very high values of the inductor Q, Q_{tk} is dominated by the capacitor quality factor. This explains the 14 GHz locking range at an inductor Q of 5000.

Figure 3.30 shows that Eq. 2.17 is only valid until a maximum inductor quality factor ($Q_{opt} = 30$), after which the locking range starts decreasing. The range that is independent of Q, which is below a Q of 30, is limited by $g_{q,max}$ and the output capacitance (it follows Eq. 2.17). This is described in Fig. 3.30 as the $g_{q,max}$-limited regime. At very low Q values, a drop in the locking range is noticed. This is due to the higher effective output capacitance, which causes a shift in the divider free-running frequency and a drop in the locking range. In the Q-limited regime, as Q gets higher, the frequency components at the edge of the locking range start decreasing in amplitude due to the reduced bandwidth. When those frequency

Table 3.4 Design values for the divider chain circuit

60 GHz dual mixing direct ILFD			30 GHz dual mixing direct ILFD		
Parameter		Value	Parameter		Value
Cross-coupled pair (Mdiv60)	M	24	Cross-coupled pair (Mdiv30)	M	48
	L	80 nm		L	80 nm
Input injecting transistors (Minj60)	M	20	Input injecting transistors (Minj30)	M	20
	L	80 nm		L	80 nm
Current mirror (Mtail60)	M	96	Current mirror (Mtail30)	M	96
	L	160 nm		L	160 nm
Differential inductor (Ldiv60)	Ldiv60	330 pH	Differential inductor (Ldiv30)	Ldiv30	1.2 nH
	Q	11		Q	15
Bias current (Idiv60)	I	4 mA	Bias current (Idiv30)	I	2 mA
First SCL divider			Second SCL divider		
Parameter		Value	Parameter		Value
Buffer transistor (Mb1)	M	12	Buffer transistor (Mb2)	M	12
	L	80 nm		L	80 nm
Cross-coupled transistor (Mcc1)	M	8	Cross-coupled transistor (Mcc2)	M	8
	L	80 nm		L	80 nm
Input transistor (Mclk1)	M	12	Input transistor (Mclk2)	M	12
	L	80 nm		L	80 nm
Current mirror (Mtail1)	M	64	Current mirror (Mtail2)	M	64
	L	160 nm		L	160 nm
Load resistor (Rlatch1)	R	700 Ω	Load resistor (Rlatch2)	R	1 kΩ
Bias current (Ilatch1)	I	2 mA	Bias current (Ilatch2)	I	1 mA

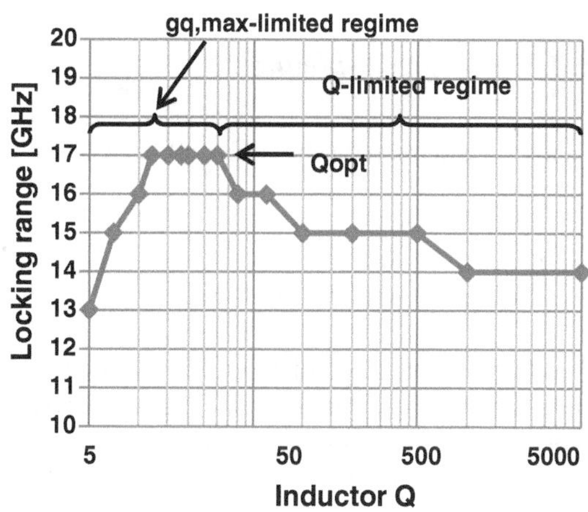

Fig. 3.30 Locking range behavior with inductor quality factor

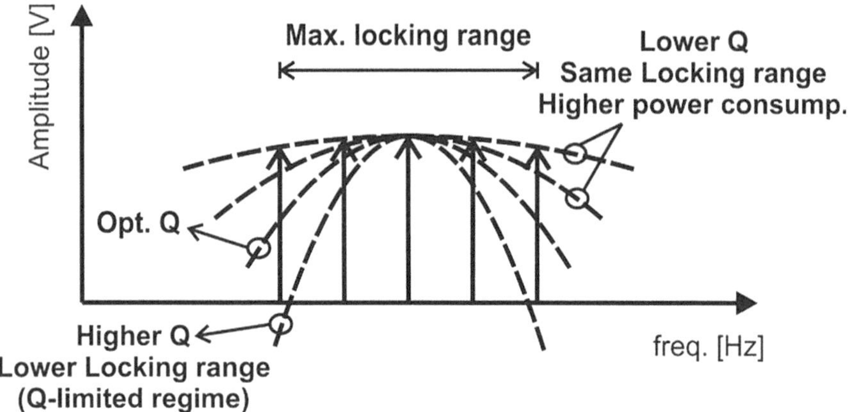

Fig. 3.31 Divider output spectrum with quality factor curves assuming equal output amplitude

components are filtered out by the low circuit bandwidth, the locking range is reduced. Higher bias current can help at this moment to increase the amplitude of the frequency components at the edge of the locking range. This leads the locking range to increase again to its value before reducing the inductor Q. Thus, reducing the inductor Q gives the same locking range with higher bias current. Figure 3.31 shows a divider expected output spectrum with different inductor Q values assuming equal output amplitudes. The optimum Q is indicated according to above understanding. The optimum tank quality factor is the highest quality factor before the locking range starts decreasing due to the limited bandwidth.

3.4.3 Design Guidelines

General guidelines for the design of each of the divider stages are discussed in this section.

3.4.3.1 First Analog Divider Stage

The first divider stage of Fig. 3.28 is a dual mixing direct ILFD (Fig. 2.26). The inductor value (Ldiv60) should be maximized as suggested by Eq. 2.17. The inductor quality factor (Qdiv60) should be maximized as suggested by Fig. 3.30. The divider can be biased with lower current at higher Qdiv60 without reducing the locking range (as long as Qdiv60 is lower than Q_{opt}). The maximum inductor value is limited by the self-resonance frequency, and the maximum quality factor is limited by the technology. The cross-coupled pair (Mdiv60) and the dual-injecting transistors (Minj60) should provide, together with the inductor value, a free-running frequency of 30 GHz. The current source transistors (Mtail60) should be increased until the drain-source voltage of the diode-connected transistor is close enough to that of the tail transistor and the current is copied with the lowest loss. For this, the length of Mtail60 can be increased two or three times of its minimum value (80 nm) to reduce the short channel effect. The tail current (Idiv60) should be increased until the locking range does not increase anymore with current ($g_{q,max}$-limited region), and the output voltage swing is high enough to drive the following stage. Finally, the output conductance of the input injecting transistors should be maximized by using different combinations of Minj60 and Mdiv60 (keeping the same free-running frequency) to get the maximum locking range.

3.4.3.2 Second Analog Divider Stage

The second divider stage uses the same topology as the first one. Thus, only steps to migrate design parameters from the 60 GHz-input divider stage to the 30 GHz divider are going to be discussed. The free-running frequency is divided by 2 (15 GHz). This means that the LC product should be multiplied by 4.

$$L_{30}C_{30} = 4L_{60}C_{60} \tag{3.6}$$

where the subscript indicates the block input frequency in GHz. The locking range of the second divider is only required to be one-half that of the 60 GHz divider.

$$\Delta\omega_{60} = \frac{2g_{q,\max,60}}{C_{60}}$$

$$\Delta\omega_{30} = \frac{\Delta\omega_{60}}{2} = \frac{g_{q,\max,60}}{C_{60}} = \frac{2g_{q,\max,30}}{C_{30}} \tag{3.7}$$

If the input amplitude of the second divider is one-half that of the 60 GHz divider, then $g_{q,max,30}$ equals to $g_{q,max,60}/2$. From Eq. 3.7, it follows that C_{30} is the same as C_{60}. By substituting in Eq. 3.6, the 30 GHz stage inductor should be 4 times larger than the 60 GHz stage one.

3.4.3.3 Static Dividers

Static dividers are used in the third and fourth divider blocks. The maximum frequency of operation determines the locking range. This is mainly affected by the total capacitance at the divider output terminals and power consumption. The output capacitance should be reduced and more power consumption should be used for higher frequency operation. Firstly, a bias current should be assumed. The tail transistors should be increased until the current is copied with the lowest loss. As explained in Sect. 2.3.2, the cross-coupled devices can be 1.5x smaller in gate width than buffer devices. Both buffer and cross-coupled transistors should be set to their minimum gate width value (keeping acceptable matching properties [8]). This is to minimize the output capacitance. The load resistance (R_D) should be chosen to maximize the output voltage swing at the specified bias current. Finally, the bias current can be changed, with all the circuit parameters re-optimized, to control the divider speed. The maximum operating frequency is designed to be slightly higher than the maximum input frequency to save in the power consumption.

3.4.4 Design Values

Final design values according to the discussed guidelines are shown in Table 3.4. The finger width used is 1 μm. M defines the number of fingers and L is the channel length.

3.4.5 Simulation Results

In the following section, the schematic simulation results of the divider chain are plotted. Each divider block is loaded by the rest of the divider chain while being tested. Ideal inductors with estimated quality factors of 11 and 15 were used in the first and second analog dividers, respectively. Additional capacitances of 10 fF each were added at the analog divider outputs to simulate layout parasitics. Transient simulations and FFT were used to collect the data.

3.4.5.1 60 GHz Divider

Figure 3.32 shows the locking range at different input levels for the first divide-by-two stage. At rail-to-rail input signal (500 mV-peak = 4 dBm), 15 GHz locking range can be achieved around 59.5 GHz. Note that locking range is reduced with lower input amplitudes. This emphasizes the importance of higher output voltage swings from the LO buffer. The minimum affordable locking range in our system is 8 GHz + process margin. This can be around 9 GHz, which can be achieved at around 0 dBm input signal (0.3 V-peak). Output power at rail-to-rail input signal is shown in Fig. 3.33. This is going to feed the following 30 GHz divider stage.

3.4.5.2 30 GHz Divider

The locking range of the second divider is shown in Fig. 3.34. A 22 GHz locking range can be achieved at rail-to-rail input (0.5 V-peak). Only 4 GHz input locking

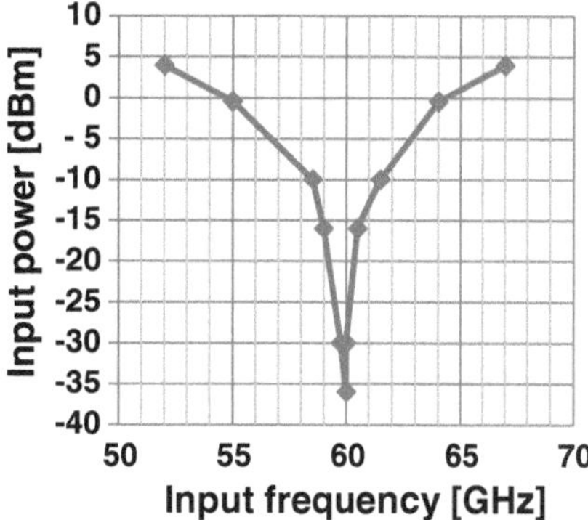

Fig. 3.32 60 GHz divider input sensitivity curve

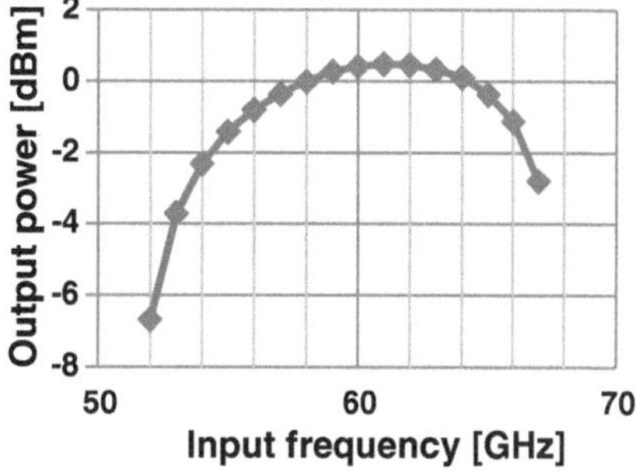

Fig. 3.33 60 GHz divider output power at 4 dBm input signal

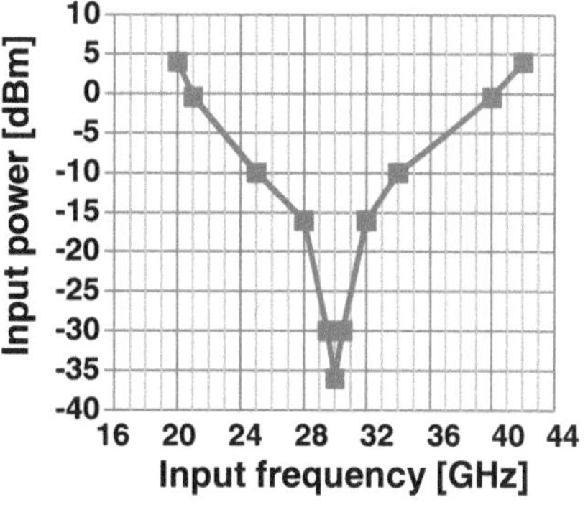

Fig. 3.34 30 GHz divider input sensitivity curve

range is required at 30 GHz to match the 8 GHz locking range of the preceding 60 GHz divider. This can be achieved with around −15 dBm input signal. This gives a good margin for the 60 GHz divider output signal that can be reduced due to lower input signals or process variations. Output power at rail-to-rail input signal is shown in Fig. 3.35. This is slightly reduced at lower input levels.

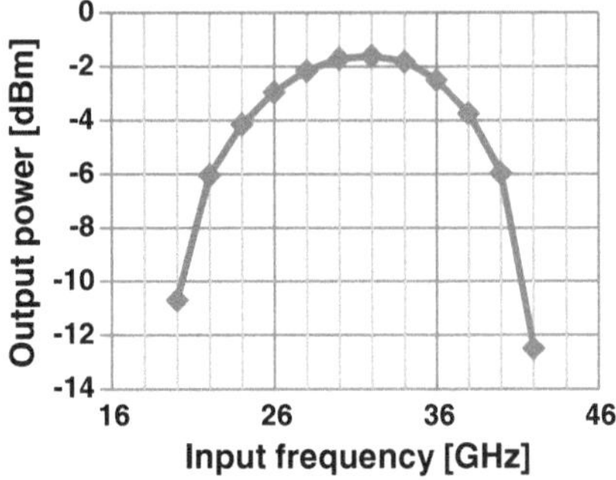

Fig. 3.35 30 GHz divider output power at 4 dBm input signal

3.4.5.3 First CML Divider

Input sensitivity curve and output power level of the first static divider are shown in Figs. 3.36 and 3.37, respectively. A maximum input frequency of 25 GHz can be achieved by the SCL divider at rail-to-rail input signal. This divider operates at 15 GHz, and only needs 2 GHz input locking range. A minimum signal of −15 dBm can be afforded because of the lower edge of the locking range.

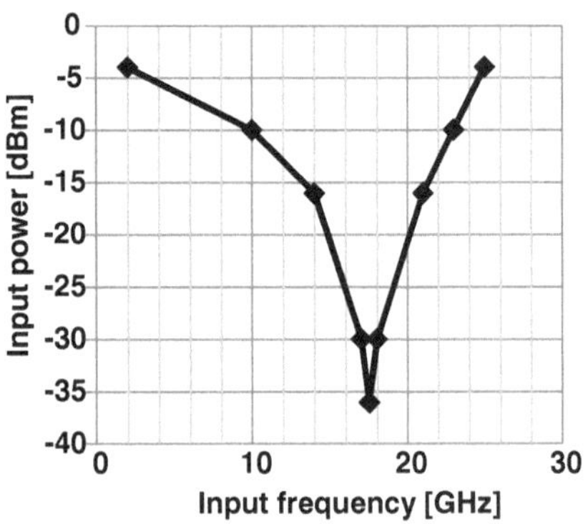

Fig. 3.36 First static divider input sensitivity curve

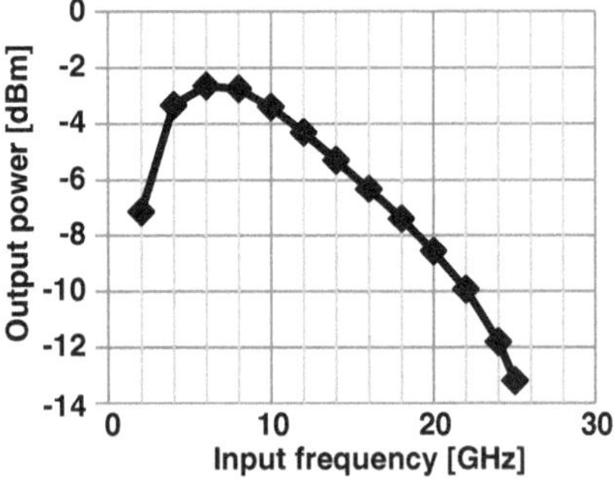

Fig. 3.37 First static divider output power at 4 dBm input signal

3.4.5.4 Second CML Divider

As shown in Fig. 3.38, maximum input frequency of the second static divider is slightly lower than that of the first one. The difference between both dividers is the lower power consumption in the second one (and the adjusted load resistance accordingly). The divider operates at 7.5 GHz and only needs 1 GHz input locking range. Minimum affordable input signal power is around −14 dBm because of the lower edge of the locking range. Output power at rail-to-rail input signal is shown in Fig. 3.39. This is going to feed the following differential to single-ended circuit that is loaded by the 50 Ω off-chip resistance through the 2 inverting buffers.

3.4.5.5 The Complete Divider Chain

With a rail-to-rail input signal to the divider chain, Fig. 3.40 shows signal power levels at different nodes. Circuit nodes at which signal levels are plotted are defined by numbers from 0 to 6. These are shown in Fig. 3.28. Signal frequencies are indicated in the plot. Figure 3.40 shows a drop in signal level at 15 GHz (after the

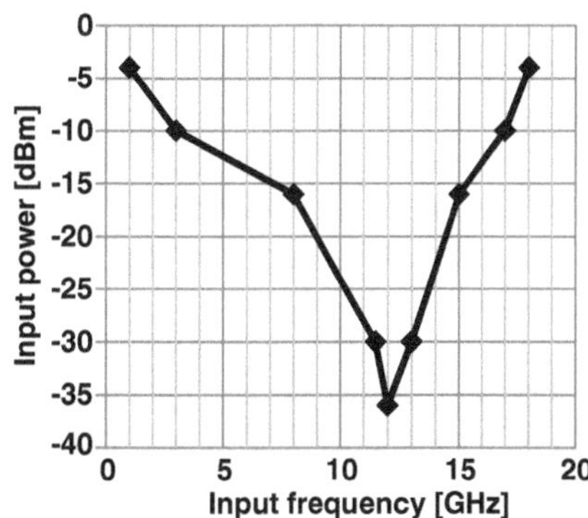

Fig. 3.38 Second static divider input sensitivity curve

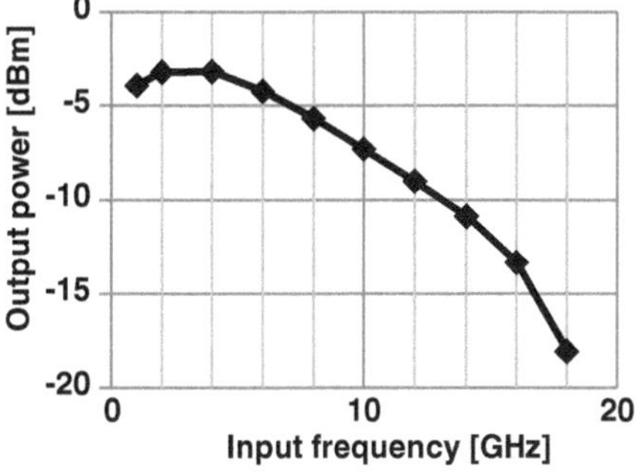

Fig. 3.39 Second static divider output power at 4 dBm input signal

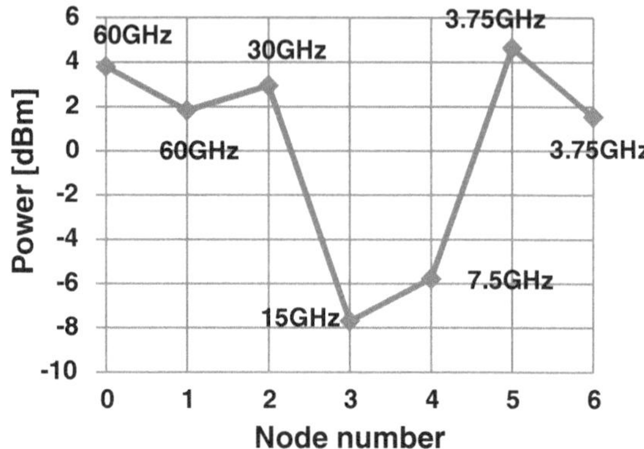

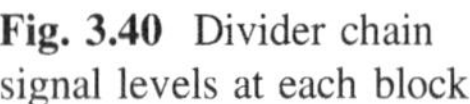
Fig. 3.40 Divider chain signal levels at each block

second divider stage). This is because the output of the 60 GHz divider is around 0.5 dBm (Fig. 3.33). This caused the output signal of the 30 GHz divider to be reduced as compared to Fig. 3.35.

3.4.6 Performance Summary

A summary of the performance of each divider block is shown in Table 3.5. The center frequency listed in the table is the frequency at which the divider locks with minimum input signal amplitude. The four divider blocks consume 9 mW. The rest of the chain, including the differential to single-ended block, and the two inverters consume DC power consumption of 4.3 mW.

In Sect. 3.4.3.2, the locking range of the 30 GHz divider was expected to be halved at one-half of the input signal (same locking range at the same input signal level). This is not the case because $g_{q,max}$ was assumed to be the same. The 30 GHz divider output has a lower bias voltage (due to the larger inductor value, and thus, higher DC resistance) and amplitude compared to the 60 GHz divider. This increases $g_{q,max}$, which results in a higher value for the locking range.

Table 3.5 Performance summary of each divider block

Divider block	Locking range at rail-to-rail input (GHz)	Center frequency (GHz)	Power consumption (mW)
60 GHz divider	52–67	60	4
30 GHz divider	20–42	30	2
First SCL divider	2–25	17.5	2
Second SCL divider	1–18	12	1

3.5 LNA and Mixer

The aim of this section is to migrate an already-existing LNA and mixer combination [13] from a 45 nm process to the 90 nm one. No architectural changes were performed on the circuit. Circuit elements including actives and passives were redesigned in the 90 nm technology. Target specifications are a maximum conversion gain of 26 dB at 60 GHz input, 6 dB noise figure, +3.5 dBm output −1 dB compression point and −12 dBm third-order intercept point (IIP3) in the high gain (HG) mode. These are the measured values for the 45 nm design.

3.5.1 Circuit Schematic

The circuit schematic is shown, reprinted from [13], in Fig. 3.41. The circuit includes a two-stage, single-ended, inductively-degenerated, common-source cascode amplifier. The second stage is loaded by a transformer, which drives a double-balanced mixer for down-conversion. The actual circuit includes two similar double-balanced mixers for I and Q signals from the LO input. In our circuit, the mixer performance was

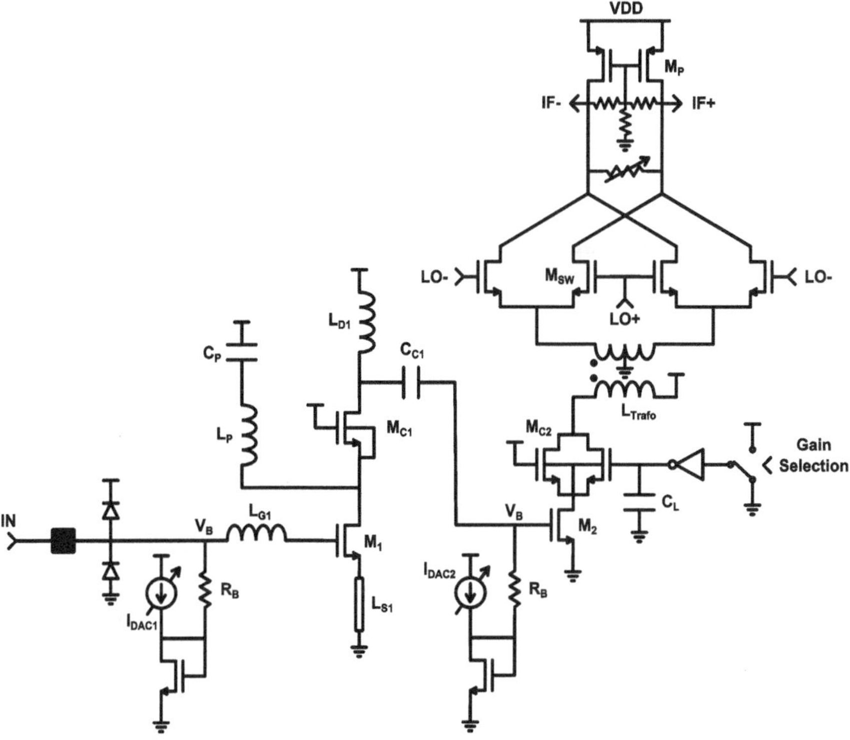

Fig. 3.41 Circuit schematic of the LNA + mixer combination [13]

checked using a resistive load instead of the diode-connected PMOS load in Fig. 3.41. This is because no information was available about the mixer load.

3.5.2 Design Guidelines

As explained in Sect. 2.4.2, gain, noise figure (NF), and input impedance matching for the LNA can be achieved through adjusting the size of transistor M1, together with inductor values L_{S1} and L_{G1}. Inductor L_P should be sized to absorb the parasitic capacitance at the cascode node. Capacitor C_P is just to prevent the cascode node from being biased, through inductor L_P, at the supply voltage. Inductor L_{D1} and the second stage transformer should provide maximum gain at the required 60 GHz frequency. The transformer should also absorb the parasitic capacitance at the source node of the mixer switches Msw. This, together with the LO buffer transformer, allow for larger switch transistors without losing the bandwidth.

3.5.3 Design Values

Design values for the two-stage LNA and mixer, excluding the bandgap reference, current DAC and digital circuitry are listed in Table 3.6. Minimum length (80 nm) is used in all circuit transistors. The supply voltage is 1 V. M defines the number of fingers and W is the channel width.

3.5.4 Simulation Results

Schematic simulation results of the LNA and mixer combination are going to be plotted in the following sub-sections. The circuit is simulated with ideal inductors (L_{S1}, L_{G1}, L_{D1} and L_P) with an estimated quality factor of 12. The transformer RLCG model is used, as shown in Fig. 3.10. An intermediate frequency of 1 GHz is used in the simulation. Estimates of the parasitic capacitances for the layout were kept from the original design.

3.5.4.1 Conversion Gain

Conversion gain as a function of LO input power is shown in Fig. 3.42 for a load resistance (RL) of 1 kΩ. Different LO bias voltages ($P_{LO,DC}$) were used to choose the operating voltage. This can be controlled at the LO buffer transformer center tap (Fig. 3.12). A mixer switch transistor should be biased close to its threshold voltage (around 0.25 V) for symmetric switching (an ideally square wave output current).

Table 3.6 Design values for the LNA and mixer

Transistor sizes		
Parameter		Value
Common-source transistor (M1, M2)	W	1.5 µm
	M	64
Cascode transistor (Mc1)	W	1.5 µm
	M	64
Cascode transistor (Mc2)	W	1.5 µm
	M	32
Switch transistor (Msw)	W	1.5 µm
	M	8
Passive elements' values		
Source inductor	L_{S1}	150 pH
Gate inductor	L_{G1}	10 pH
Drain inductor	L_{D1}	40 pH
Cascode inductor	L_P	80 pH
Cascode capacitor	C_P	370 fF
Coupling capacitor	C_{C1}	370 fF
Biasing resistor	R_B	6.5 kΩ
Coupling transformer (primary = sec.)	Ltrafo	90 pH
Controlling parameters		
CS transistor bias voltage @ max. current	Vb	660 mV
LO bias voltage	$V_{LO,DC}$	200 mV

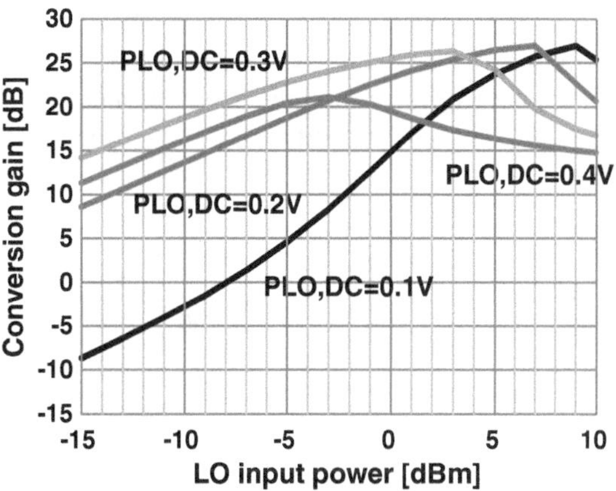

Fig. 3.42 Conversion gain versus LO input power for RL = 1 kΩ

That's why the gain is decreased at LO bias voltages of 0.1 and 0.4 V. The conversion gain is increased with the input LO amplitude until the mixer switch transistors go into the triode region. At a $P_{LO,DC}$ of 0.2 V, the mixer switch transistors go into the triode region at 7 dBm (compared to 3 dBm at $P_{LO,DC}$ of 0.3 V)

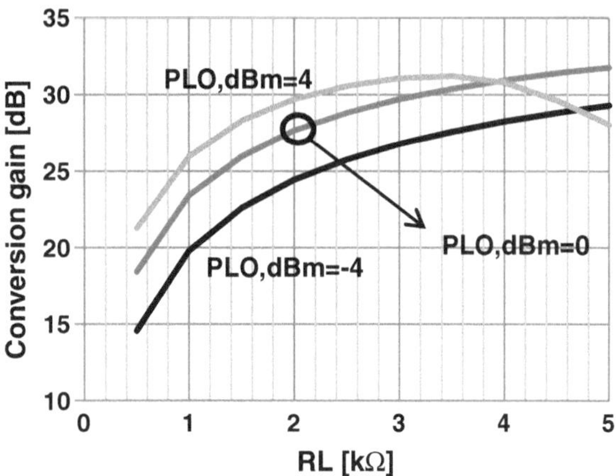

Fig. 3.43 Conversion gain versus mixer load resistance for different LO input power values

LO input power ($P_{LO,dBm}$), and the conversion gain is still large enough at smaller LO input signals. So, a LO bias voltage of 200 mV was chosen for our operation.

Conversion gain versus load resistance at different LO input power levels is shown in Fig. 3.43. Conversion gain is proportional to the load resistance. However, higher resistance values causes larger voltage drops that reduce headroom as the drain voltage of the mixer transistor is biased at a lower voltage. With larger input swing at the LO input, the mixer transistor can easily go into the triode regime, causing the conversion gain to drop with LO input amplitude. That's why the gain is dropped at 3.5 kΩ for $P_{LO,dBm}$ of 4 dBm while it keeps increasing for lower LO input power values (expected to drop at higher load resistance values).

Conversion gain at 5 kΩ load resistance is shown in Fig. 3.44. Gain starts decreasing at a lower value for LO input power compared to case with 1 kΩ (Fig. 3.42). Thus, operation with a 5 kΩ resistive load is only advised at low input LO power values (below 3 dBm).

Conversion gain across the input frequency band is shown in Fig. 3.45. Maximum conversion gain of 26.3 dB at 60 GHz is achieved at 1 kΩ load resistance, 200 mV LO

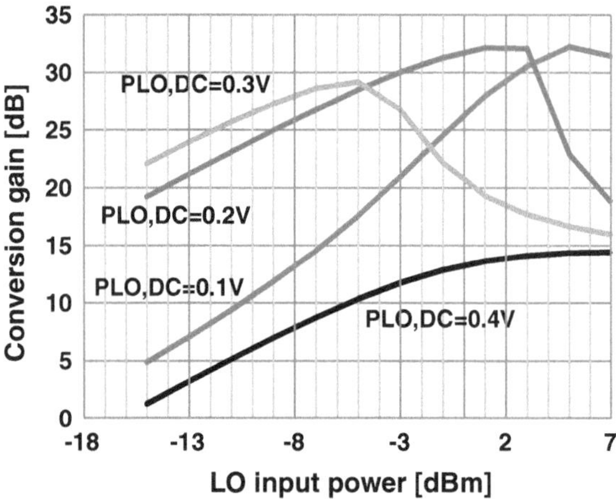

Fig. 3.44 Conversion gain versus LO input power for RL = 5 kΩ

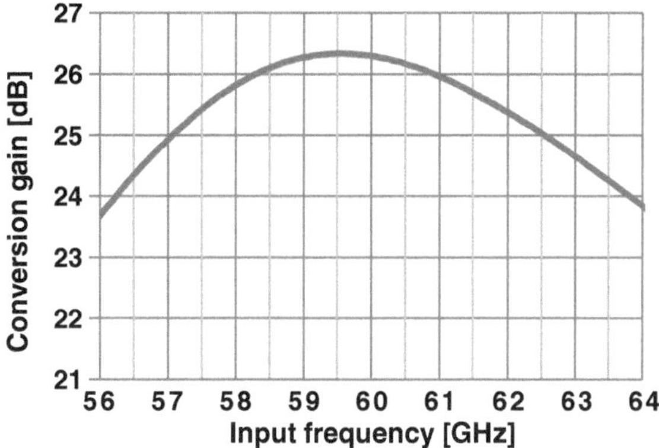

Fig. 3.45 Conversion gain versus RF input frequency for RL = 1 kΩ and $P_{LO,dBm}$ = 4 dBm

bias voltage, and 4 dBm LO input power. The circuit doesn't provide a constant gain over the 7 GHz input bandwidth. This explains the need for wide-band LNA design if channel bonding is used in the system and the entire bandwidth is used.

3.5.4.2 Noise Figure and S11

Double-side band noise figure is going to be used for the following NF results. Noise figure and input return loss are shown in Fig. 3.46a, b, respectively. The noise figure is 5 dB and S11 is lower than −10 dB over the whole 7 GHz input frequency band.

3.5.4.3 Linearity

Output −1 dB compression point and third order intercept point are shown in Fig. 3.47. The load resistance is set to 1 kΩ, and the LO input power is 4 dBm. Output P-1 dB from simulation is +5.8 dBm and IIP3 is −9.7 dBm.

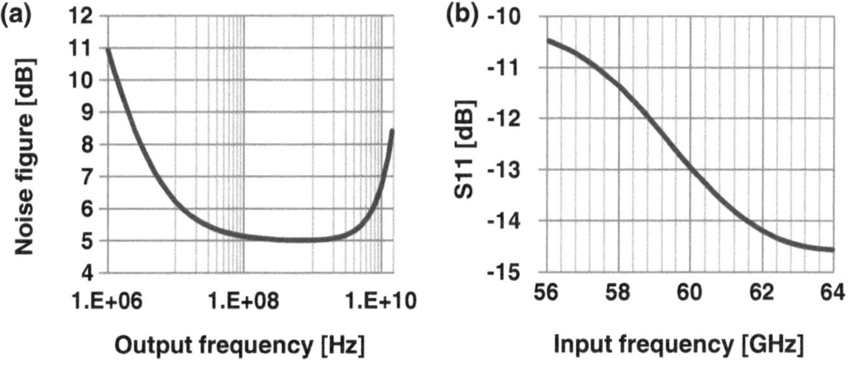

Fig. 3.46 **a** Noise figure versus IF output frequency, and **b** S11 versus RF input frequency

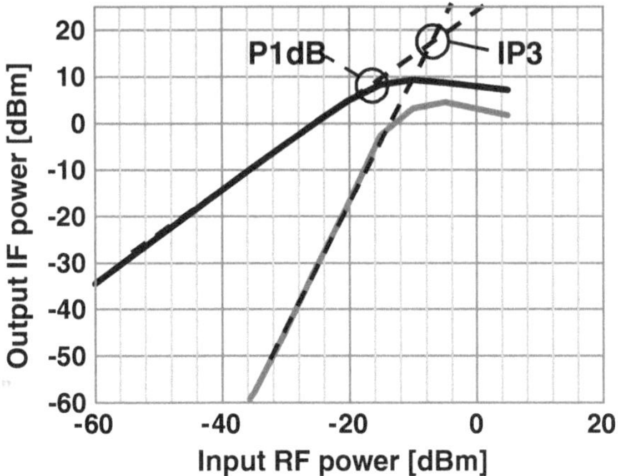

Fig. 3.47 −1 dB compression point and third order intercept point for RL = 1 kΩ

The circuit consumes 41.6 mW from 1 V supply. This is because of the increased LNA transistor sizes during migration. The biasing circuit, on the other hand, is kept without changes.

3.5.4.4 Results at Reduced Power Consumption

If we switch off one bit of the current DAC, we can get a reduced bias voltage for the LNA CS transistors (VGS = 500 mV). This reduces the total power consumption to 21.8 mW. The gain could, however, be increased by using a higher load resistance. At a load resistance of 1.5 kΩ, conversion gain versus LO input power is shown in Fig. 3.48. The conversion gain starts to drop at 6 dBm input LO power (0.63 V-peak). A conversion gain of 26.77 dB can be achieved at 4 dBm

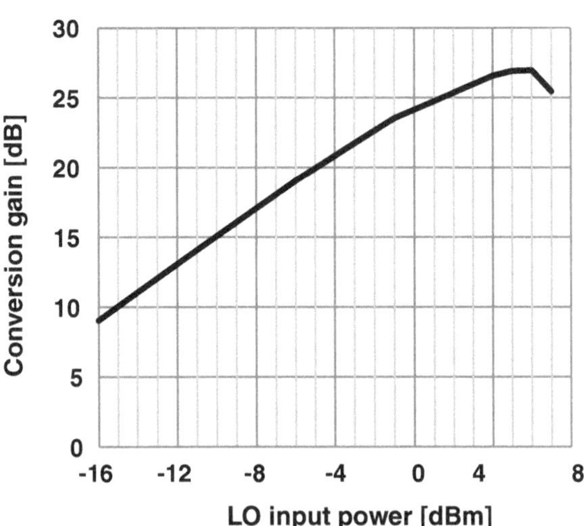

Fig. 3.48 Conversion gain versus LO input power at 21.8 mW total power consumption

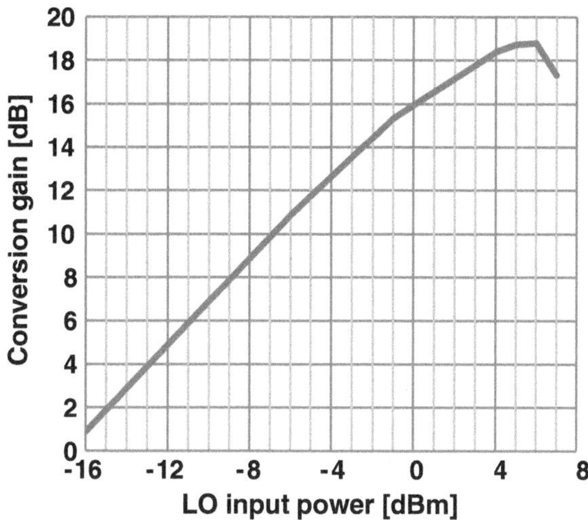

Fig. 3.49 Conversion gain versus LO input power at 10.55 mW total power consumption

Table 3.7 Performance summary of the LNA and mixer combination

Parameter	Required	Achieved (RP)
Conversion gain (CG) (dB)	26	26.77
Noise figure (NF) (dB)	6	5.88
Output −1 dB compression point (dBm)	+3.5	+6.3
Input-referred 3rd order intercept point (IIP3) (dBm)	−12	−8.6
Max. input reflection coefficient (S11) (dB)	−10	−10
Total power consumption (mW)	23	21.8

input power (0.5 V-peak). S11 is below 12.5 dB for the entire frequency range (57–64 GHz). NF is 5.88 dB at 1 GHz IF frequency. Output −1 dB compression point is 6.3 dBm and IIP3 is −8.6 dBm. These results are very close to the high power (HP) mode results with a 50 % improvement in the power consumption.

At minimum power consumption (LP mode), the LNA CS transistors are biased at a gate-source voltage of 380 mV. Total power consumption is then 10.55 mW. Conversion gain for a mixer load resistance of 1.5 kΩ is shown in Fig. 3.49 versus LO input power. A conversion gain of 17.7 dB can be achieved at 4 dBm LO input power (0.5 V-peak). S11 is below −15 dB over the 57–64 GHz input frequency range, and NF is 10.11 dB. Output −1 dB compression point is +5.8 dBm and IIP3 is −2.4 dBm.

3.5.5 Performance Summary

A comparison between the required specifications and the design simulations in the reduced power (RP) mode is shown in Table 3.7. The reduced power mode has very close results to the high power mode except for the improved power consumption. Thus, it should be used instead.

References

1. Advanced Design System (ADS), Agilent Technologies. [Online]. http://www.agilent.com/find/eesof-ads
2. Scheir K (2009) CMOS building blocks for 60 GHz phased-array receivers. Ph.D. thesis, Vrije Universiteit Brussel
3. Mazzanti A, Andreani P (2008) Class-C harmonic CMOS VCOs, with a general result on phase noise. IEEE J Solid-State Circuits 43(12):2716–2729
4. Rofougaran A et al (1998) A single-chip 900 MHz spread-spectrum wireless transceiver in 1-mm CMOS—part I: architecture and transmitter design. IEEE J Solid-State Circuits 33 (4):515–534
5. Wahl P (2009) Exploring the pathways of faster optical telecommunication in 90 nm CMOS: design, analysis and layout of a quadrature voltage-controlled oscillator. M.Sc. thesis, Vrije Universiteit Brussel
6. Kral A, Behbahani F, Abidi AA (1998) RF-CMOS oscillators with switched tuning. In: Proceedings of the IEEE custom integrated circuits conference, Santa Clara, CA, pp 555–558, May 1998
7. Hegazi E, Abidi AA (2003) Varactor characteristics, oscillator tuning curves, and AM–FM conversion. IEEE J Solid-State Circuits 38(6)
8. Pelgrom MJM, Duinmaijer ACJ, Welbers APG (1989) Matching properties of MOS transistors. IEEE J Solid-State Circuits 24(5):1433–1440
9. Berny AD, Niknejad AM, Meyer RG (2005) A 1.8-GHz LC VCO with 1.3-GHz tuning range and digital amplitude calibration. IEEE J Solid-State Circuits 40(4):909–917
10. Levantino S et al (2002) Frequency dependence on bias current in 5-GHz CMOS VCOs: impact on tuning range and flicker noise upconversion. IEEE J Solid-State Circuits 37 (8):1003–1011
11. Chan WL, Long JR, Spirito M, Pekarik JJ (2009) A 60 GHz band 1 V 11.5 dBm power amplifier with 11 % PAE in 65 nm CMOS. In: IEEE ISSCC digest of technical papers, pp 380–381a
12. Wu C-Y, Yu C-Y (2007) Design and analysis of a millimeter-wave direct injection-locked frequency divider with large frequency locking range. IEEE Trans Microw Theory Tech 55 (8):1649–1658
13. Borremans J, Raczkowski K, Wambacq P (2009) A digitally controlled compact 57-to-66 GHz front-end in 45 nm digital CMOS. In: IEEE ISSCC digest of technical papers, pp 492–493a

Chapter 4
Top-Level Design

After all the front-end circuit blocks are discussed, the top-level schematic design is going to be presented in this chapter. The circuit performance after putting blocks together is expected to differ from the performance of the blocks separately. This is due to the effect of actual loading of one block by another, as compared to the expected loading while dealing with each section alone.

4.1 Complete Top-Level Circuit

In our top-level, we still have another challenge. This is to let the oscillator drive both the mixer and the divider. Two transformer-coupled CS amplifiers provide suitable voltage swings to both the mixer and divider inputs. The top-level circuit now includes the QVCO, LO buffers, divider chain, LNA and mixer. Top-level schematic and simulation results are shown in the following sections.

4.1.1 Circuit Schematic

Figure 4.1 shows the top-level schematic of all of the blocks connected together. The P-QVCO, the divider chain, the LNA and mixer from Sects. 3.3, 3.4 and 3.5 respectively, are used. Ideal transformers with an estimated quality factor and coupling coefficient of 8 and 0.8, respectively, are used in the simulation.

4.1.2 Design Choices

The divider input transistor is 40 μm wide and the mixer input transistor is 12 μm wide. A single CS differential stage was not enough to buffer both loads. So, two

K. Khalaf et al., *Data Transmission at Millimeter Waves*,
Lecture Notes in Electrical Engineering 346, DOI 10.1007/978-3-662-46938-5_4

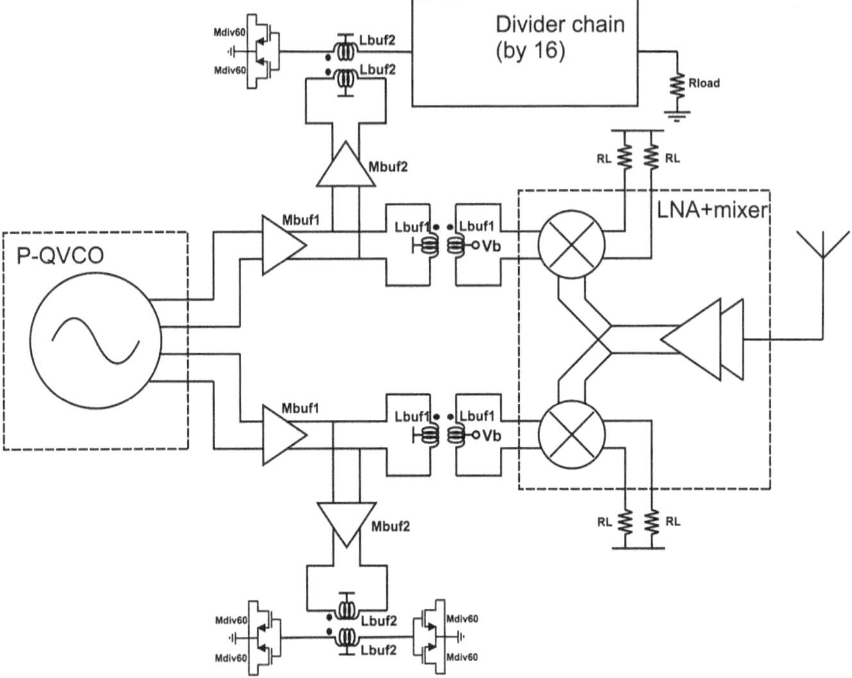

Fig. 4.1 Front-end top-level circuit schematic

buffer stages (M_{buf1} and M_{buf2} in Fig. 4.1) are used instead. One choice was to load the first buffer with the second buffer and the second buffer with both the mixer and divider loads. The mixer circuit can operate (with enough gain) at lower input amplitude than the divider. As shown in Fig. 3.32, the divider locking range is increased from around 6–15 GHz if the input amplitude is increased from 0.2 to 0.5 V-peak. Meanwhile, Fig. 3.48 shows a reduction of around 2 dBm in mixer conversion gain for the same change in the input amplitude. Thus, instead of providing the mixer and divider circuit with the same input amplitude, one buffer stage is chosen to buffer the mixer circuit and two for the divider. The buffer configuration is shown in Fig. 4.1. This allows more swing to be delivered to the divider input (without an effective degradation in the mixer performance) as compared to the first choice (two buffers loading both mixer and divider circuits). Downsides of the additional buffer are the added chip area and power consumption. However, an approximately 2 × 50 μm × 50 μm additional chip area (dominated by the differential inductor size) and 2 × 10 mA additional power consumption are not so significant in a system with an estimated chip area of 0.13 mm^2 and power consumption of 92 mA.

4.1.3 Design Values

The P-QVCO values are listed in Table 3.2, where parameters related to the P-QVCO don't depend on the load. The divider chain values are listed in Table 3.4. The reduced power LNA is used in the top-level. The LNA and mixer design values are shown in Table 3.6. The bias voltage of the CS node in the LNA is 500 mV. The load resistance is 1.5 kΩ, and the LO input bias voltage (Vb) is 200 mV. These are the same values used in the reduced power version of the LNA and mixer combination. The transformer values are chosen to tune out the parasitic capacitance and maximize the buffer gain at 60 GHz. The gate width of the buffer transistor is a trade-off between buffer gain and input impedance. Higher buffer gate width causes a shift in the oscillator operating frequencies due to the higher load capacitance. Design values for the two LO buffers are shown in Table 4.1. The transistor finger width (W) of all the buffer transistors is 1 μm. M defines the number of fingers and L is the channel length.

4.1.4 Simulation Results

Transient simulations of the circuit are used to estimate the QVCO output amplitude. As shown in Fig. 4.2, an approximate value of 0.22 V-peak is achieved at the QVCO output. This is 50 mV-peak lower than the value achieved before (Fig. 3.18), because different circuit loading the buffer was used. The buffer load affects the QVCO output through feedback via gate-drain parasitic capacitance of the buffer transistor (Cgd, buf). The oscillation frequency is expected to change as well. As shown in Fig. 4.3, output tuning range of 56–64 GHz is achieved. This is 800 MHz

Table 4.1 LO buffers design values for the complete top-level circuit

Parameter		Value
Transistor sizes		
First buffer transistor (M_{buf1})	M	14
	L	80 nm
Second buffer transistor (M_{buf2})	M	20
	L	80 nm
Current mirror transistors (Mcm, buf)	M	208
	L	300 nm
Passive elements' values		
First buffer transformer (primary = sec.)	L_{buf1}	200 pH
Second buffer transformer (primary = sec.)	L_{buf2}	190 pH
Controlling parameters		
First buffer bias current	I_{buf1}	5 mA
Second buffer bias current	I_{buf2}	10 mA

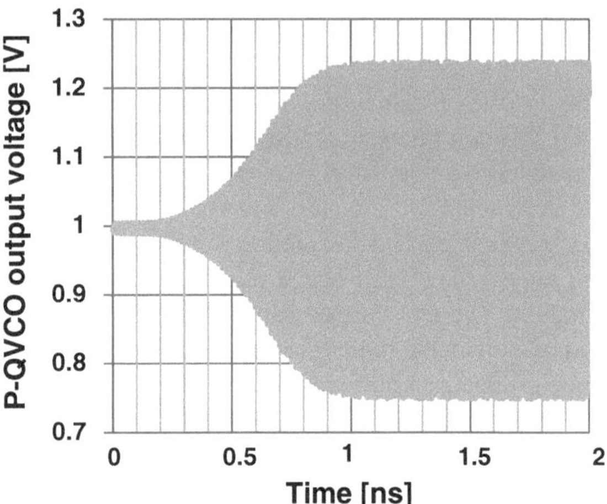

Fig. 4.2 P-QVCO transient output voltage in the complete top-level system

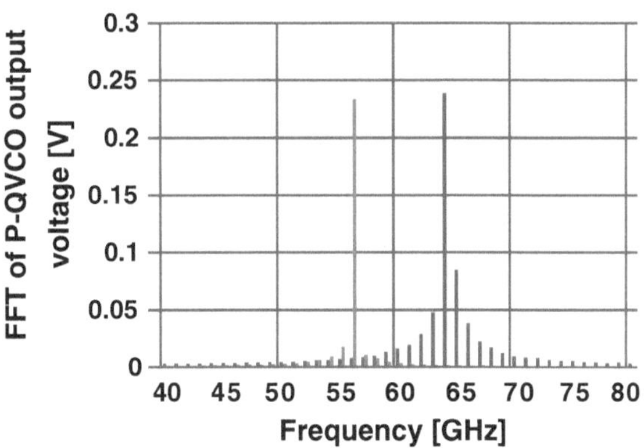

Fig. 4.3 FFT of the P-QVCO output at max. and min. tuning ranges

lower than the simulated 56.8–64.8 GHz in Fig. 3.18, but still within the required 8 GHz range. Note that the spectral width of the QVCO output at higher frequency indicates phase noise degradation, which is expected from the previous QVCO simulations (see Fig. 3.19). With a QVCO single-ended output of 200 mV, voltages at the divider and mixer inputs are shown in Fig. 4.4 with different Lbuf1 values. A 0.473 V-peak (0.946 V-pp) signal is achieved at the divider input. This is close to the rail-to-rail input swing required for the maximum locking range of the divider (Fig. 3.32). At the mixer input, however, a value of 330 mV-peak is achieved. With an increased load resistance (RL = 2 kΩ instead of 1.5 kΩ), a conversion gain of 26 dB is still achievable. The whole front-end consumes 26.6 mW from the QVCO, 13.3 mW from the divider chain, 21.8 mW from the LNA and mixer combination, 2 × 5 mW from the first buffer and 2 × 10 mW from the second buffer. This gives a total of 91.7 mW for the front-end circuit.

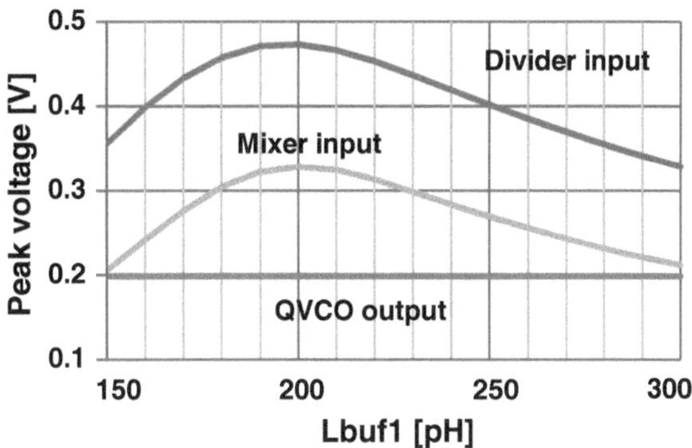

Fig. 4.4 Voltage levels at the mixer and divider inputs versus L_{buf1} value

4.2 QVCO and Divider Sub-system

Due to time limitations, the whole front-end top-level is not laid-out. So, a QVCO, LO buffer, first stage divider and an output buffer are to be considered for layout. The top-level schematic of this sub-system and its simulation results are going to be presented in this section.

4.2.1 *Circuit Schematic*

The top-level schematic of the sub-system is shown in Fig. 4.5. Compared to Fig. 3.12, one buffer output drives the 60 GHz divide-by-2 stage.

4.2.2 *Design Choices*

As shown previously in Fig. 3.18, rail-to-rail output voltage swing was not achieved when the LO buffer is loaded with two divider injecting transistors, each with 20 μm gate width. As explained in Sect. 2.3.1, using two injecting transistors can increase the injecting transistor output conductance width the drawback of larger input device. Thus, a single injecting transistor is implemented (with a width of 24 μm), which provides a smaller load for the LO buffer. As will be shown in the simulation results (Fig. 4.8), this causes a reduced locking range compared to Fig. 3.32, but can still be matched to the QVCO output tuning range (with a divider input voltage swing of 0.4 V). Using two buffers to drive the divider is recommended for future work.

As explained in Sect. 3.3.3.10, a choice can be made between the transformer-coupled and the inductively-tuned buffer configurations depending on the buffer

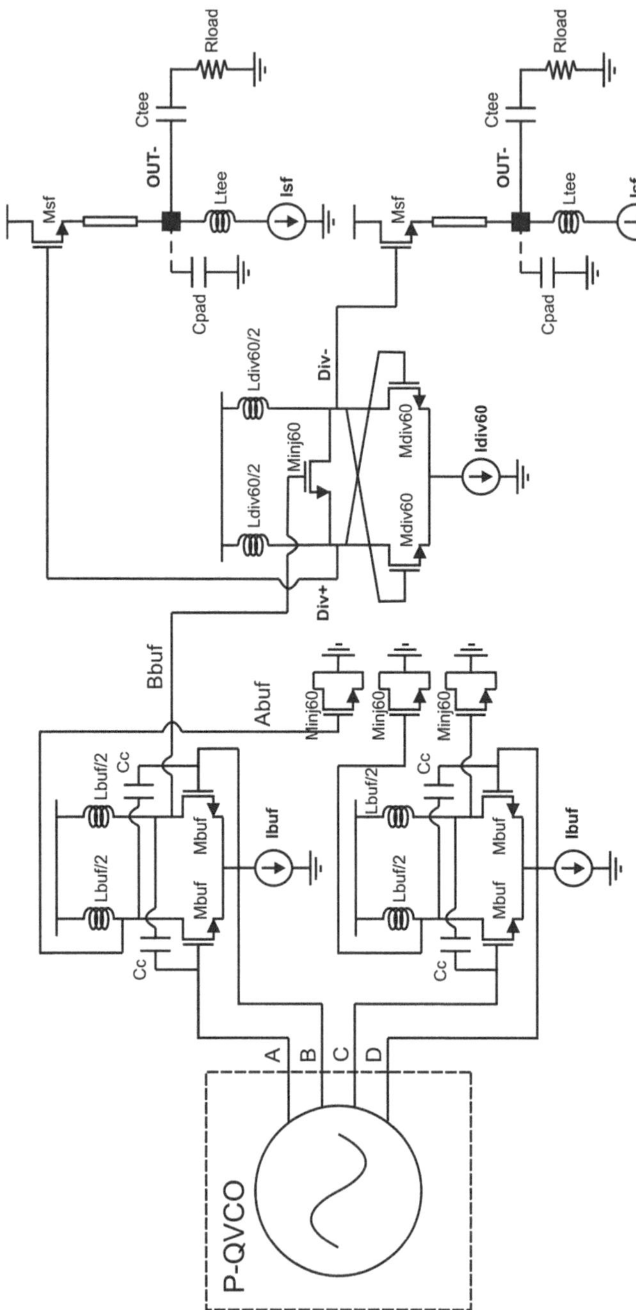

Fig. 4.5 QVCO and divider sub-system circuit schematic

load resistance. Both configurations were simulated with the single injecting transistor divider. The buffer output voltage swing was higher in the case of inductively-tuned buffer. This is because the resistive contribution of the 24 μm wide divider input to the buffer output resistance is higher than that of the transformer (see Eqs. 2.11, 2.14 and 2.16).

Neutralization capacitors (Cc) are used in the LO buffer to provide more stability [1]. They cancel the effect of the feedback parasitic capacitance (Cgd, buf) and help increase the oscillator output amplitude. The neutralization capacitors are implemented as MOS transistors with the gate and source connected together. This adds a gate-source and bulk capacitances to the gate-drain capacitance required for neutralization. Hence, the neutralization capacitors can be designed smaller than the buffer transistor to account for these additional parasitic capacitances. The phase noise improvement due to the neutralization capacitors will be shown in Sect. 4.2.4.3.

The output buffer is a source follower with its source connected to the output pads through 50 Ω transmission lines. The source-follower is biased with an external current source (Isf). Ltee is an external inductor that is used to increase the current source output impedance and Ctee is an external decoupling capacitor. Together with the source follower size, Isf adjusts the transistor output impedance to 50 Ω.

4.2.3 Design Values

Design values of the P-QVCO are the same as those listed in Table 3.2. Design values of the LO buffer, divider and output buffer are listed in Table 4.2. The transistor finger width of all the following transistors is 1 μm. M defines the number of fingers and L is the channel length.

4.2.4 Simulation Results

The simulation results of each circuit block will be presented. Each block is loaded by the following stage during simulation.

4.2.4.1 Source Follower

The real part of the source follower output impedance (Zsf, out) is shown in Fig. 4.6. Figure 4.6a shows the variation of Re(Zsf, out) with different values of the source-follower width (Msf) if Ltee is connected to ground. At Msf = 16 and Isf connected, output impedance of 50 Ω can be achieved at 4.5 mA as shown in Fig. 4.6b. Lower current can be achieved at larger Msf (e.g., 1.3 mA at Msf of

Table 4.2 Design values of the sub-system blocks (without the QVCO)

Parameter		Value
Transistor sizes		
LO buffer transistor (M_{buf})	M	14
	L	80 nm
Neutralization transistor (Mn)	M	12
	L	80 nm
LO buffer current mirror (Mcm, buf)	M	208
	L	300 nm
Divider transistor (Mdiv60)	M	26
	L	80 nm
Injecting transistor (Minj60)	M	24
	L	80 nm
Divider current mirror (Mcm, div60)	M	96
	L	160 nm
Output buffer transistor (Msf)	M	16
	L	80 nm
Passive elements' values		
LO buffer inductor (without Cc)	L_{buf}	300 pH
LO buffer inductor (with Cc)	L_{buf}	160 pH
Divider inductor	Ldiv60	300 pH
Controlling parameters		
LO buffer bias current	I_{buf}	10 mA
Divider bias current	Idiv60	3 mA
Output buffer bias current	Isf	5 mA

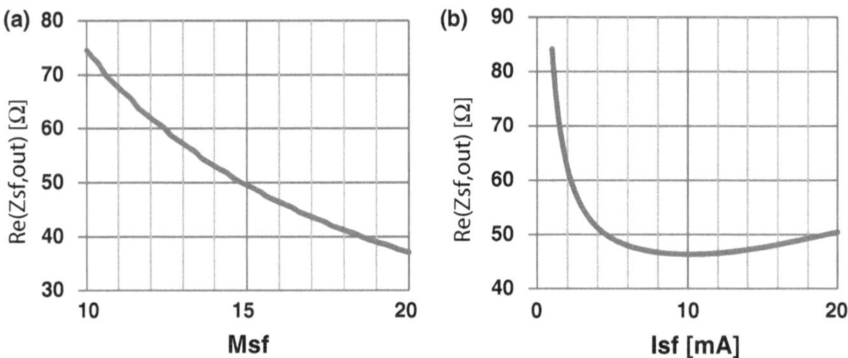

Fig. 4.6 Real part of the source follower output impedance **a** without current source, and **b** with current source at Msf = 16

30 μm), but this will load the divider and degrade the locking range. The small-signal gain of the source-follower at a bias current of 4.5 mA is shown versus frequency in Fig. 4.7a. A value of 0.41× is achieved for the buffer gain at 30 GHz.

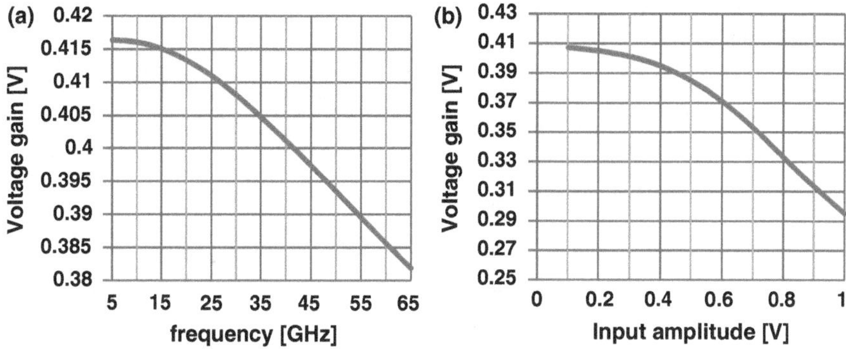

Fig. 4.7 **a** Small- and **b** large-signal voltage gain of the source follower

The large-signal gain versus the input amplitude is shown in Fig. 4.7b. A value of around 0dBm ($\sim$0.32 V) is expected from the divider output. The large signal gain at this value is still above 0.4×.

4.2.4.2 Divider

The input sensitivity curve of the 60 GHz divider and the divider output power versus input frequency is shown in Figs. 4.8 and 4.9, respectively. At 2 dBm ($\sim$0.4 V-peak) input, 10 GHz locking range is predicted from simulations at the schematic level. The inductor used is an ideal one with estimated quality factor of 15. The inductor is implemented with the following parameter values: N = 2, W = 3 μm, S = 2 μm and OD = 55 μm.

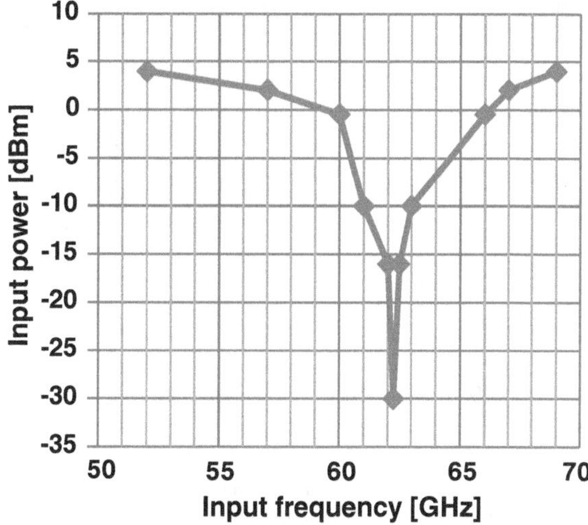

Fig. 4.8 Input sensitivity curve of the sub-system divider

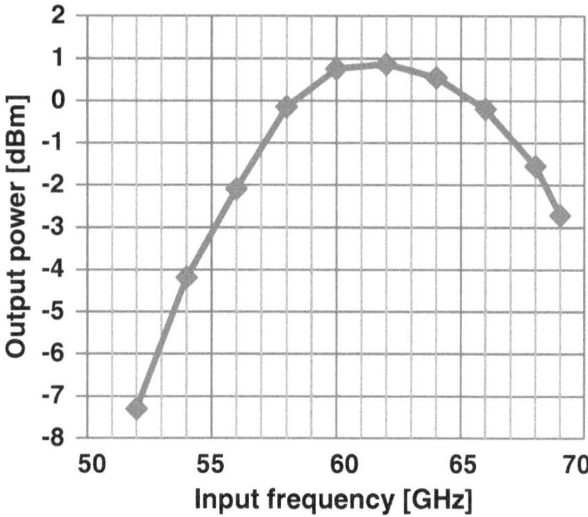

Fig. 4.9 Output power of the sub-system divider

4.2.4.3 QVCO and LO Buffer

The output voltage levels and phase noise results are going to be shown across the tuning range in the following sections. The results of the circuit with and without the neutralization capacitors are presented.

The performance of the sub-system without adding neutralization capacitors is shown in Fig. 4.10. The divider input (B_{buf} as indicated in Fig. 4.5) can reach about 0.4 V-peak, and the tuning range is 8.4 GHz. The phase noise is increased at higher frequencies of the tuning range. This is noticed before in Sect. 3.3.5.2. Neutralization capacitors are going to be used in the following section to cancel the effect of Cgd, buf. This improves the oscillator output, and phase noise accordingly (Eq. 2.9).

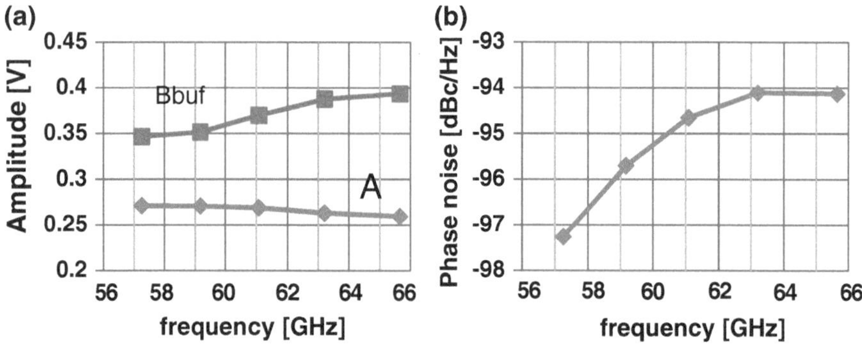

Fig. 4.10 **a** Amplitudes and **b** phase noise of the sub-system QVCO and LO buffer versus tuning range without neutralization capacitors

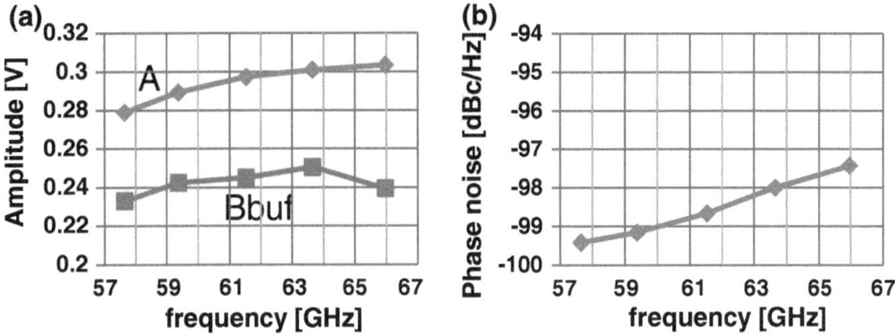

Fig. 4.11 **a** Amplitudes and **b** phase noise of the sub-system QVCO and LO buffer versus tuning range with neutralization capacitors

Figure 4.11 shows the performance of the QVCO and LO buffer after using the neutralization capacitors. The phase noise shows a noticeable improvement due to the increased oscillator amplitude (A). A maximum of −97.4 dBc/Hz is predicted from simulation at maximum operating frequency. All parameters are the same as the case without Cc except for the buffer inductor, which is 160 pH instead of the 300 pH. This is due to the additional capacitance seen at the buffer output node. The gain is also reduced, and a maximum of 0.25 V-peak is simulated at the divider input (B_{buf}). The circuit consumes a total of 48.6 mW. The output voltage can be increased by increasing the bias current. This causes more power to be consumed in the QVCO.

The previous discussion shows a trade-off between output amplitude (or power consumption) and phase noise. The neutralization capacitors reduce total phase noise, but also the LO buffer gain. As the main target of the QVCO design is to achieve low phase noise, the circuit using neutralization capacitors is chosen to be laid-out and tested. This will be shown in the next chapter.

Reference

1. Chan WL, Long JR, Spirito M, Pekarik JJ (2009) A 60 GHz band 1 V 11.5 dBm power amplifier with 11 % PAE in 65 nm CMOS. In: IEEE ISSCC Digest of Technical Papers, pp. 380–381a

Chapter 5
Layout and Post-layout Simulations

The layout of the QVCO buffer sub-system of Sect. 4.2 is discussed in this chapter. Post-layout simulation results are also provided.

5.1 Physical Layout

The schematic used is shown in Fig. 4.5. The circuit consists of a P-QVCO, LO buffer, a divide-by-two stage and output buffers. Current sources are provided externally. The top-level circuit layout with the pads connecting input-output signals (IO ring) is shown in Fig. 5.1. The IO ring doesn't enclose the circuit core symmetrically due to the available chip area. The circuit core is only 0.33 mm × 0.2 mm, and the chip with IO ring is 1.1 mm × 0.63 mm. Unit elements (5 μm × 5 μm) of technology metal stack are used to form a low-resistance ground plane. Ground cells are used in the chip to connect the external supply and ground pads to the internal nodes.

Figure 5.2 is a zoom-into the circuit core. The QVCO inductor (Lvco) parameters are mentioned in Sect. 3.3.3.3. An inductance value of 39.6 pH and a quality factor of 14.2 are predicted using this inductor. The LO buffer and divider inductors (L_{buf} and L_{div60}) parameters are listed in Table 5.1. The interconnect lines to the active elements are taken into account.

Figure 5.3 shows the QVCO layout without the ground cells. The buffer transistor (M_{buf}) is placed close to the QVCO output. The QVCO layout, without the current mirror transistors and biasing resistors (R_d), is symmetrical around its center.

5.2 Nominal Simulation Result

The following simulations are performed with typical process parameters, a supply voltage of 1 V and at a temperature of 27 °C.

K. Khalaf et al., *Data Transmission at Millimeter Waves*,
Lecture Notes in Electrical Engineering 346, DOI 10.1007/978-3-662-46938-5_5

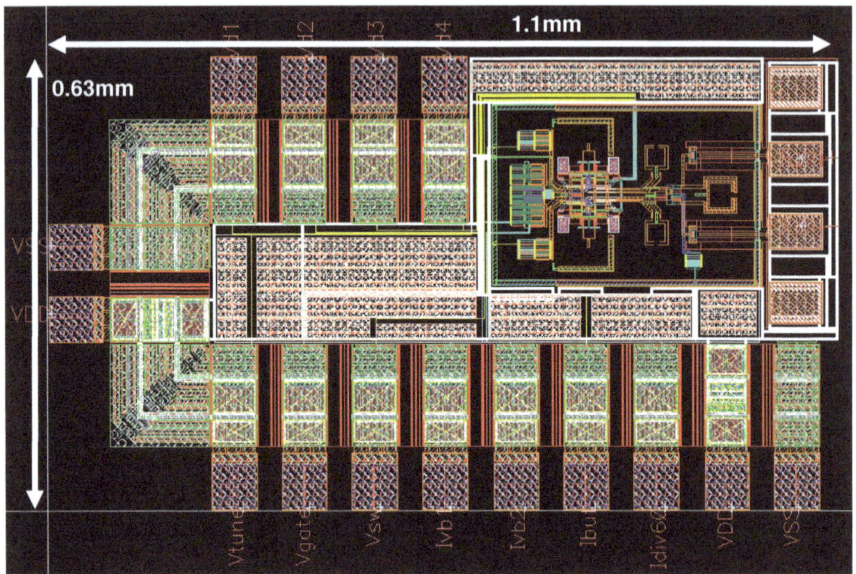

Fig. 5.1 Top-level circuit layout with IO ring

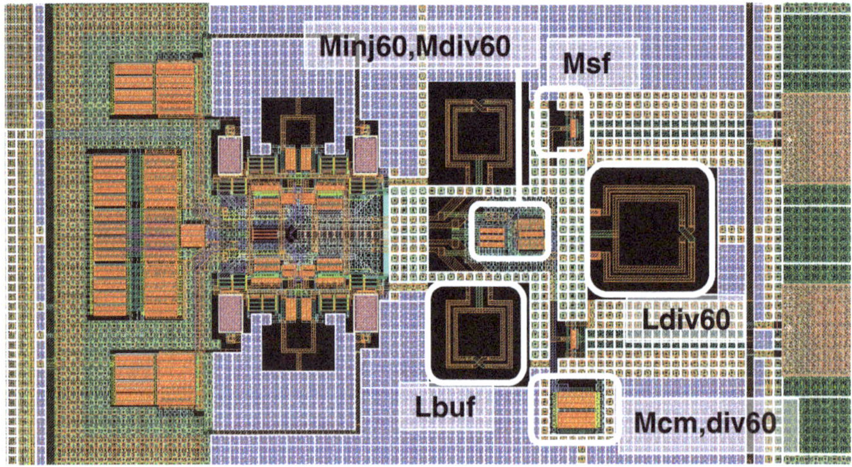

Fig. 5.2 The circuit core including the QVCO, LO buffer, divider and output buffer

Table 5.1 Inductor parameters for the LO buffer and divider in the test chip

Parameter	N	W (μm)	S (μm)	OD (μm)	L (pH)	Q	fo (GHz)
L_{buf}	2	2	2	35	240	10	60
L_{div60}	2	3	2	48	247	14.9	30

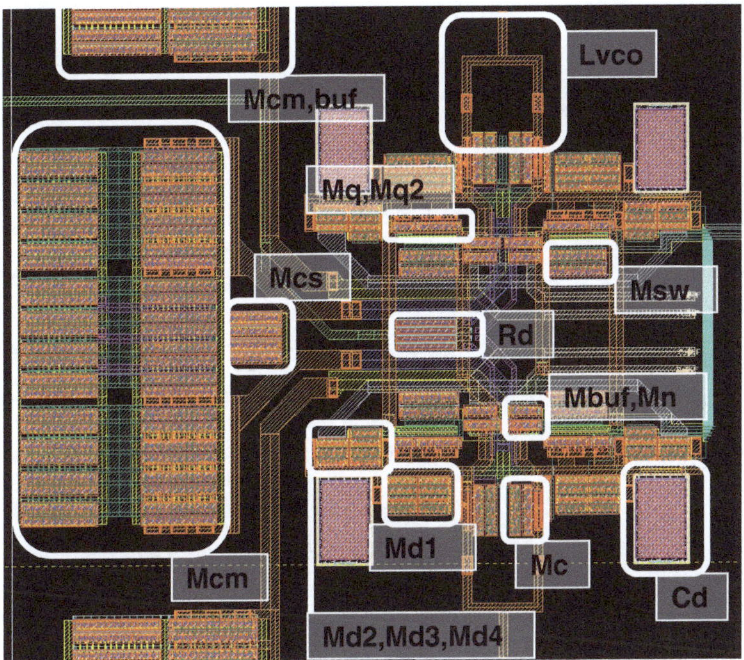

Fig. 5.3 The P-QVCO, and LO buffer active elements with the ground cells omitted

5.2.1 Divider

The aim of the divider stage is to translate the output signal from 60 to 30 GHz. Thus, the free-running frequency should be around 30 GHz. As shown in Table 4.2, a value of 300 pH was selected for the divider inductor (Ldiv60). This can be implemented with an inductor of OD of 55 μm. An inductor with OD of 48 μm is, mistakenly, chosen instead. This inductor has 300 pH at 60 GHz, but it only has 247 pH at 30 GHz. As shown in Figs. 5.4 and 5.5, a shift of 8 GHz is caused in the input locking range (at minimum input amplitude) when the OD = 48 μm inductor is used. This can cause a mismatch between the QVCO output frequency range and the divider input locking range, and the divider output will not track the input frequency anymore.

Table 5.2 Worst case PVT simulation results

	Process corner	V_{DD} (V)	Temp. (°C)	Phase noise (dBc/Hz)	QVCO output (A) (mV)	Divider input (B_{buf}) (mV)	Oscillation frequency (GHz)
Schematic	FF	1.1	0	−99.93	382.4	257.1	59.46
	SS	0.9	100	−87.67	212.3	186.9	62.7
Layout	FF	1.1	0	−98.13	460	261.2	56.32
	SS	0.9	100	×	×	×	×

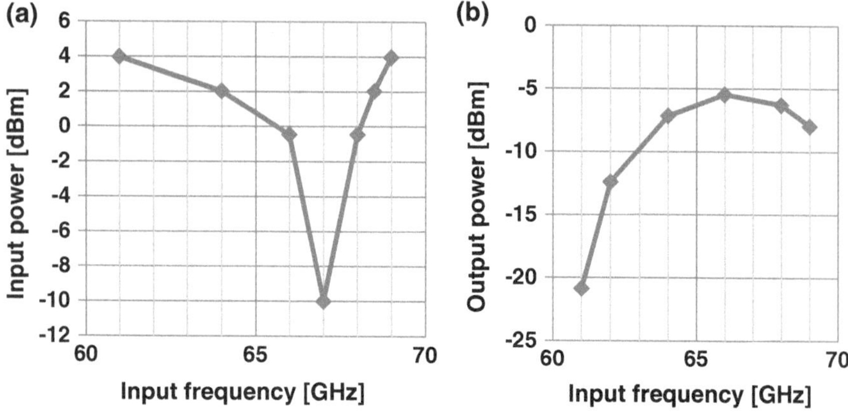

Fig. 5.4 Divider **a** input sensitivity curve and **b** output power for a 48 μm OD inductor

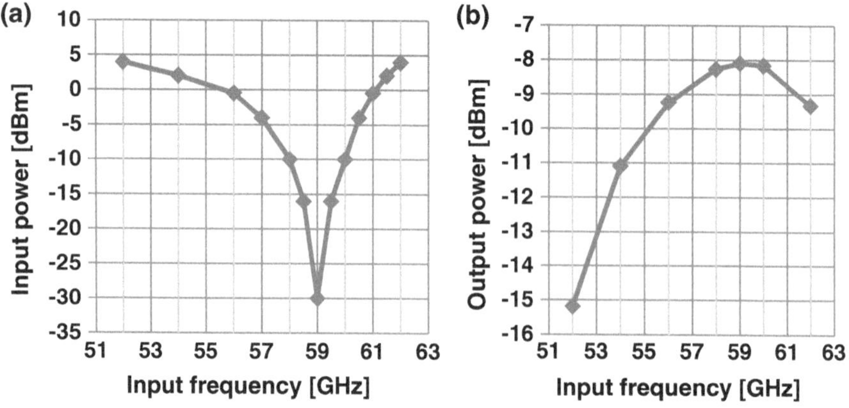

Fig. 5.5 Divider **a** input sensitivity curve and **b** output power for a 55 μm OD inductor

5.2.2 *QVCO and LO Buffer*

The post-layout simulation results after the P-QVCO and LO buffer are shown in this section. The divider inductor used in these simulations is the one with OD of 55 μm. Figure 5.6 shows the output amplitudes and phase noise versus the QVCO bias current source (Ivco) at the middle of the tuning range and a gate voltage of 1 V. More current is required to start the oscillation as compared to the schematic because of the parasitic resistance. Also the inductor value is around 40 pH, which is lower than the one used in the schematic. This is to overcome the additional layout parasitic capacitance. The reduced QVCO inductor leads to lower amplitude, and thus more current is required for the start-up. With a QVCO bias current of

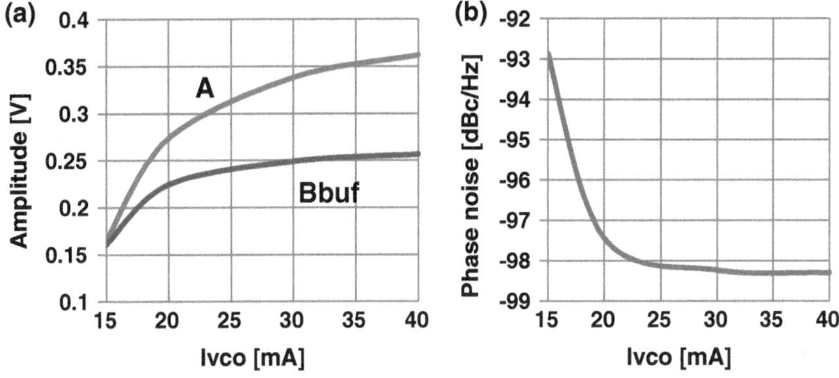

Fig. 5.6 Variation with QVCO bias current at V_{gate} of 1 V

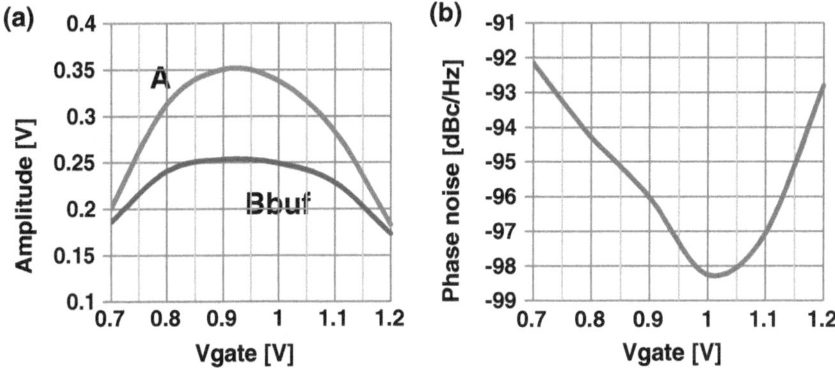

Fig. 5.7 Variations with gate voltage at Ivco of 30 mA

30 mA, the gate voltage (V_{gate}) is varied in Fig. 5.7, and phase noise reaches a minimum value at V_{gate} of 1 V. Figure 5.8 shows the QVCO and LO buffer amplitudes and phase noise at 30 mA bias current and 1 V_{gate} voltage. The simulated LO buffer output is around 0.25 V and the tuning range is 6.3 GHz centered at 58.5 GHz. A maximum phase noise value of −97.4 dBc/Hz is predicted from simulation, which is the main target specification. The LO buffer output amplitude can be increased either by removing the neutralization capacitors and increasing the buffer transistor (M_{buf}) gate-width or using a two-stage buffer to boost the voltage swing. The first solution increases the phase noise and the second solution costs more supply current and chip area.

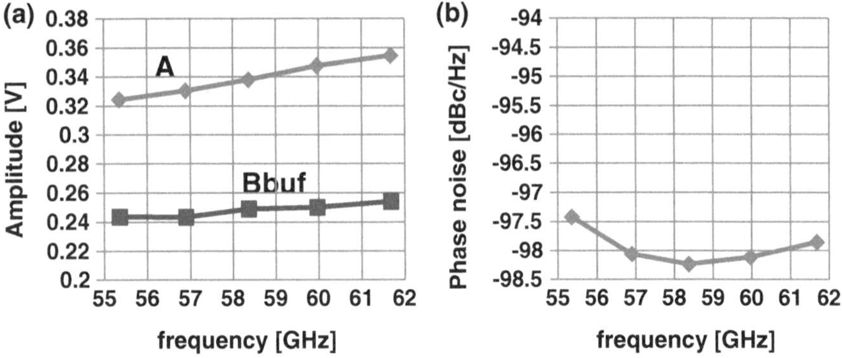

Fig. 5.8 Performance at V_{gate} of 1 V and Ivco of 30 mA over the tuning range

5.3 PVT Simulations

The aim of this section is to provide simulation results at different process corners, supply voltages and operating temperatures (PV and T). As shown in Fig. 5.9, the power supply is swept from 0.9 to 1.1 V (±10 % with a nominal value of 1 V) using a typical process with a temperature of 27 °C (room temperature). Simulations with different process corners are performed first on the schematic level. MOS process corners are: typical (TT), slow-NMOS-slow-PMOS (SS), slow-NMOS-fast-PMOS (SF), fast-NMOS-slow-PMOS (FS) and fast-NMOS-fast-PMOS (FF). Figure 5.10 shows the output amplitudes, phase noise and oscillation frequency versus process corners at nominal supply voltage (1 V) and room temperature (27 °C). The QVCO output amplitude at the slow process (SS) is the smallest. Higher temperature leads to lower carrier mobility and slower operation. Thus, the worst case is to simulate a

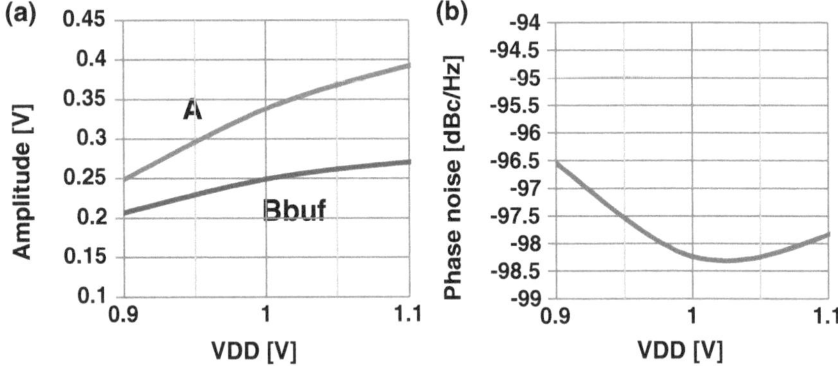

Fig. 5.9 Amplitude and phase noise variations of the QVCO and LO buffer with supply voltage at V_{gate} of 1 V and Ivco of 30 mA

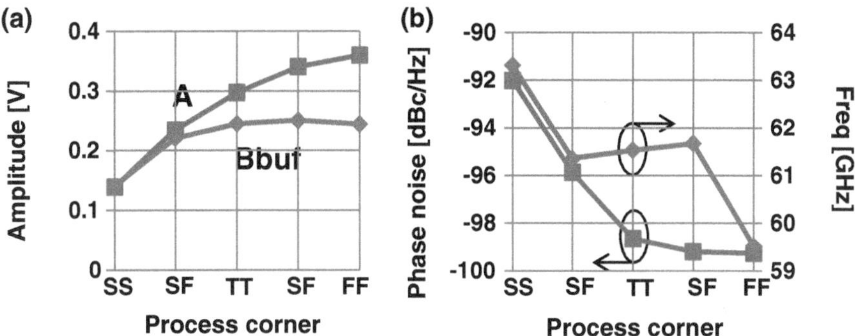

Fig. 5.10 **a** Amplitudes, **b** phase noise and oscillation frequency schematic simulations versus process corners at 27 °C

slow process with high temperature (e.g., 100 °C) and low supply voltage (0.9 V), and a fast process with low temperature (e.g., 0 °C) and high supply voltage (1.1 V). Worst case results are summarized in Table 5.2 for pre- and post-layout simulations. The slow combination needs more current for start-up. The listed schematic simulation results for PVT of SS, 0.9 V and 100 °C use a gate voltage of 0.7 V and a QVCO bias current of 25 mA (the default values are a gate voltage of 0.6 V and bias current of 10 mA). The post-layout worst case slow simulation did not start up.

Chapter 6
Conclusions

In this work, a receiver front-end at 60 GHz is explored. The circuit includes a QVCO, divider chain, LNA, mixer and LO buffers. A test-chip including the QVCO, LO buffer and the first stage divider is designed to verify the key components of the receiver design. The summary and recommendations for future work are discussed in this chapter.

6.1 Summary

Chapter 1 is an introduction, including the 60 GHz band standards and system architecture. The unlicensed frequency band between 57 GHz and 66 GHz is assigned for 60 GHz operation in different countries. This can be used in applications that need data rates up to tens of gigabits per second according to the IEEE 802.15.3c and ECMA-387 standards, such as short range cable replacement with very high speed wireless links.

In Chap. 2, a theoretical background on the blocks used in the circuit is provided. The cross-coupled LC VCO theory is elaborated with more details on phase noise. The parallel, series and gate-modulated QVCO topologies are presented. For the LO buffer, the inductively-tuned and transformer-coupled common-source configurations were analyzed. The ILFD theory for use in the first and second divider blocks is presented. The other two divider blocks use static high frequency CML divide-by-two circuits. The chapter ends with noise figure and linearity background with the used inductively-degenerated cascode LNA and active mixer.

The schematics, design steps and simulation results of the circuit blocks are included in Chap. 3. The parallel QVCO is gate-decoupled and biased externally to achieve a better phase noise performance. Variable coupling was also implemented to ensure oscillator start-up. The P-QVCO achieved phase noise values of −98.9 to 94.8 dBc/Hz over the 8 GHz tuning range. Voltage levels of up to 0.52 V-peak were achieved at the LO buffer output. The bottom-series QVCO achieved approximately the same performance, but with 10 mW power consumption compared to the 26. 6 mW consumed by the P-QVCO. Two ILFD circuits were designed at 60 and

K. Khalaf et al., *Data Transmission at Millimeter Waves*,
Lecture Notes in Electrical Engineering 346, DOI 10.1007/978-3-662-46938-5_6

30 GHz. More insight into the effect of the inductor quality factor of the locking range was introduced. Two regions of operation were defined according to the value of the quality factor: the Q-limited and gq,max-limited regimes. In the gq,max-limited region, the locking range doesn't depend on the inductor quality factor as long as the injecting transistor output conductance and total capacitance are fixed. Locking ranges up to 15 GHz are achieved at 60 GHz using the dual-mixing ILFD. A maximum operating frequency of 25 GHz is achieved using the static CML divider. The whole divider chain consumes 9 mW without the output buffers. The LNA and mixer combination achieves a maximum conversion gain of 26.77 dB and a noise figure of 5.88 dB. The output −1 dB compression point is +6.3dBm, IIP3 is −8.6dBm and it consumes 21.8 mW including all biasing circuitry.

In Chap. 4, the top-level design of the receiver front-end is presented. Two LO buffers were used to drive the mixer and divider chain. The P-QVCO, LO buffer and the first stage of the divider chain are connected in a separate sub-system. With the use of neutralization capacitors in the LO buffer, maximum phase noise of −97. 4 dBc/Hz is predicted from simulations.

The layout of the test-chip is shown in Chap. 5. The post layout simulations show a maximum of −97.4dBc/Hz of phase noise and a 55.4–61.7 GHz tuning range. The LO buffer output voltage is around 0.25 V-peak. This can be improved by post-layout optimization of circuit parameters or different circuit topologies as will be discussed in the future work.

6.2 Future Work

The gate-decoupled P-QVCO is used in the test circuit. However, as shown in Sect. 3.3.6, the BS-QVCO consumes less power and gives approximately the same phase noise performance. Thus, the BS-QVCO should be utilized for lower power consumption in the QVCO. The GM-QVCO could also be explored as it is expected to provide better phase noise [1]. The QVCO inductor can, thus, be increased allowing for better modeling and lower power consumption.

The divider locking range depends on the input amplitude. Two buffer stages can be used to provide amplitudes high enough for large divider locking range. This is used in the front-end top level and can also be utilized in the test chip.

Traditional inductively-degenerated cascode LNA and active mixer are used in the design. Passive mixers could be investigated for better linearity, with the challenge of switching speed and Op-Amp bandwidth. Innovative topologies for wideband LNA should also be investigated in the future. The current LNA can handle an input bandwidth of around 2 GHz that accounts for a single channel. If channel bonding is used, more bandwidth (up to 9 GHz for 4 channels) will be required. Thus, more effort should be spent on designing a wideband LNA at 60 GHz.

Reference

1. K.-W. Cheng, et al (2009) A gate-modulated CMOS LC quadrature VCO. Radio frequency integrated circuit symposium, pp 267–270, IEEE

Transformer-Coupled Buffer

The current-output transformer-coupled commons-source differential amplifier (Fig. 2.20b) will be analyzed in this appendix. The buffer output impedance due to the transformer load (Eq. 2.15) is derived. For simplicity, the load is assumed to be only capacitive ($C_{load} = C$). We'll assume that the buffer side of the transformer is the primary side (with "p" subscript) and the load represents the secondary side (with "s" subscript). The transformer primary and secondary voltages can be written as following:

$$v_p = j\omega L\ i_p + j\omega M\ i_s \tag{A.1}$$

$$v_s = j\omega L\ i_s + j\omega M\ i_p \tag{A.2}$$

The load is only capacitive. So,

$$Z_s = \frac{v_s}{i_s} = -\frac{1}{j\omega C}$$

By substituting into A.2:

$$-\frac{i_s}{j\omega C} = j\omega L\ i_s + j\omega M\ i_p$$

$$-i_s\left(j\omega L + \frac{1}{j\omega C}\right) = j\omega M\ i_p$$

$$\frac{i_s}{i_p} = \frac{-j\omega M}{j\omega L + \frac{1}{j\omega C}} = \frac{-j\omega kL \times j\omega C}{-\omega^2 LC + 1} = \frac{\omega^2 LCk}{1 - \omega^2 LC} \tag{A.3}$$

The primary impedance from A.1 is:

$$Z_p = \frac{v_p}{i_p} = j\omega L + j\omega M \frac{i_s}{i_p}$$

K. Khalaf et al., *Data Transmission at Millimeter Waves*,
Lecture Notes in Electrical Engineering 346, DOI 10.1007/978-3-662-46938-5

Using A.3, the buffer output impedance can be derived.

$$Z_{out} = Z_p = j\omega L + j\omega kL \frac{\omega^2 LCk}{1 - \omega^2 LC}$$

$$Z_{out} = j\omega\left(L + \frac{\omega^2 L^2 C_{load} k^2}{1 - \omega^2 LC}\right)$$

Zeitfracht Medien GmbH
Ferdinand-Jühlke-Straße 7
99095 Erfurt, Deutschland
produktsicherheit@kolibri360.de